The Animal Connection

The Animal Connection

A New Perspective on What Makes Us Human

PAT SHIPMAN

W. W. NORTON & COMPANY
New York   London

For information about permission to reproduce
selections from this book, write to Permissions,
W. W. Norton & Company, Inc.,
500 Fifth Avenue, New York, NY 10110

For information about special discounts for bulk purchases, please contact
W. W. Norton Special Sales at specialsales@wwnorton.com or 800-233-4830

Manufacturing by Courier Westford
Book design by Helene Berinsky
Production manager: Devon Zahn

Library of Congress Cataloging-in-Publication Data

Shipman, Pat, 1949–
The animal connection : a new perspective on what
makes us human / Pat Shipman. — 1st ed.
    p. cm.
Includes bibliographical references and index.
ISBN 978-0-393-07054-5 (hardcover)
1. Human evolution. 2. Human-animal relationships.
3. Domestication. 4. Prehistoric peoples. I. Title.
GN281.S453 2011
304.2'7—dc22

                                2010054062

W. W. Norton & Company, Inc.
500 Fifth Avenue, New York, N.Y. 10110
www.wwnorton.com

W. W. Norton & Company Ltd.
Castle House, 75/76 Wells Street, London W1T 3QT

1 2 3 4 5 6 7 8 9 0

To all of the animals who make life worth living

Contents

Prologue

"Once you find something, how do you know if it's human?" the woman in the back of the auditorium asked.

Questions like this come up fairly often when you are a paleo-anthropologist, as I am, and give public lectures, which I do. When I get questions from nonspecialists, I try to listen with my "teacher's ear," the one that hears not what the questioner literally said but what the questioner meant. This particular question usually means, "If you find the fossil of a higher primate, how do you tell if it's an ape or a human?" The question is particularly pointed because many people know that humans and chimps have more than 99 percent of their genetic codes in common.

One answer is: "You know, fossils don't come with labels on them. You have to do as much or more work to figure out what adaptations a fossil species had as you did to find the fossil in the first place. You compare the anatomical features in the fossil with those of apes, of humans, and of other fossils to see which is the best match, which is the closest resemblance. And you hope you are going to find more fossils from the same species, so you'll have more bones to work with."

Something in the woman's voice or her intelligent eyes suggested she meant something more universal, something bigger. I
believed she was asking a very profound question indeed: What
does it mean to be human?

Perhaps surprisingly, the answer is not: "To be like us."
When you take the long perspective, *Homo sapiens sapiens* of
today is a highly evolved member of a species that has been
around and changing for hundreds of thousands of years. It may
seem paradoxical, but we were not "us" from the very beginning
of our species.

If you take both a long and broad perspective, humans are
creatures that belonged to the same genus we do, *Homo,* and
they have been around for millions of years. During that time,
our ancient relatives evolved from "them"—something not us—
into "us." But it didn't happen right away or fast.

As I started to give my questioner a detailed answer that
spelled out the unique anatomical and behavioral attributes
of humans—teeth like this, relatively big brains, bipedal
upright posture, complex tools, and fully developed language—
something flashed through my head, unbidden. It was a sort
of cinematic collage of people of the world I had seen or read
about. There were noisy, vibrant city dwellers in Prague, Accra,
or Beijing and silent, black-haired Amazonian hunter-gatherers
following narrow trails in the jungle. There were the elegant,
proud, pastoralists of arid Kenya, like the Masai in whose territory I had excavated, and solidly built, pale-skinned, down-to-
earth farmers in the American Midwest. There were tall, tanned
California girls and sari-clad beauties with dark eyes and hair
from Delhi, contrasting with the neatly built Efe Pygmies in the

dense forests of central Africa. I thought of tribal Papua New Guineans with their lush gardens, muscular Polynesians rowing their slender, efficient boats, the very dark-skinned aboriginal cowboys of Australia, and slender Javanese with their rice paddies and beloved water buffalo. Add to that turbaned Tuaregs in the Sahara, Mongolian horse peoples on the steppes, and the Inuit hunters of Alaska, clad in bulky skin parkas and leggings against the frigid cold . . . Images spun through my mind.

As I watched this mental world tour, I marveled at the diversity of human lifestyles, dwellings, body builds, coloring, habits, and habitats. I also saw, for the first time, something else that unites humankind, a behavior to which few other anthropologists or biologists have paid attention.

Included in every one of those images of human beings that flashed through my head were animals. Everywhere you go, if you find people, you find animals. I hadn't realized that before. Animals may be casual or even unwanted visitors to human habitations, treasured pets, precious livestock, pursued prey, or animate transport for people and goods alike, but they are there. We adopt them, feed them, nurture them, play with them, breed them, train them, use them, and eat them. We think with animals.

I don't know why I was surprised. I have always lived with animals myself—mostly cats and horses—and feel life is incomplete without them. Living with, communicating with, and working with animals is a great joy to me.

In that moment, I realized that living with animals is a uniquely human trait. In the wild, no other mammal lives intimately with another species. No zebras adopt antelopes, no deer looks after a baby wolf in hopes of training it to protect the herd, no monkeys bring back little mongooses to feed and pet. Except in conditions of captivity, no other animal species regularly initiates long-term nurturing relationships with individuals of another species. Tick birds feed on the parasites lodged in rhino skin and rhinos tolerate their presence, but there is no evidence that one tick bird "belongs" to a particular rhino, or vice versa. The relationship is generalized and happenstance, not one-on-one.

Behavioral universals are very hard to come by in anthropology because human behavior is so culturally conditioned and so variable. But living with animals is indeed a human characteristic and it is both very old and profoundly important. When anthropologists field questions about what makes humans human, they do not say "Our connection to animals," but I think they should.

The serious question that woman asked suddenly showed me that probably the longest and most enduring trend in human evolution has been a gradual intensification of our involvement with animals. The connection between humans and animals has in large measure defined who and what we are as a species. And, I would speculate, this connection with animals has been selected for genetically.

As I thought through the story of human evolution with this insight in mind, I realized I had misinterpreted a number of facts and milestones in human evolution. What was now clear to me was that the connection between animals and humans runs through the last 2.6 million years of our evolution like a fast-moving river, carrying our behavior and abilities into modern humanness. It was the animal connection that brought together apparently disparate developments of human evolution into one continuous theme.

Ideas multiplied in my mind in the weeks after my lecture, and I knew I had seen a new perspective on human evolution. I don't know who that person in the audience was or why she came to my lecture, but I owe her a debt of gratitude, because her honest curiosity triggered my recognition of a new trait characteristic of the human species—one with an importance that had been overlooked before. From that flash of insight came my determination to explore the implications of the human connection with animals. From that idea came this book and many new questions.

Did our connection to animals literally put the "human" into "humanity"? Are animals still central to the very essence of being human?

Do humans spend enormous amounts of money on pets and domestic animals, on trips to zoos, wildlife parks, or wilderness areas because we crave and need animal contact?

Are we approaching or living in a post-animal world—and what does that mean for our future as a species?

Questions to consider

I WANT TO STATE my hypothesis clearly at the outset.

I believe that a defining trait of the human species has been a connection with animals that has intensified in importance since at least the onset of stone toolmaking some 2.6 million years ago. Defining traits are what make humans human, what makes us different from all other animals, and they are partially or wholly encoded in our genes. I don't claim that the animal connection is the only defining trait—I will spend much of this book discussing others—but I do claim that our connection to animals is so deep, so old, and so fundamental that you really can't understand human evolution and nature without taking it into account.

These are general statements of my conviction, but they aren't scientifically testable hypotheses. To persuade a critical audience—a thinking audience—I need to spell out the implications of my hypothesis, make some predictions, run some tests, and see if those predictions are borne out by solid evidence. Arguing loudly and fervently simply isn't enough in science!

In most sciences, you'd dream up some experiment in order to test your ideas. But in paleoanthropology, you can't do experiments—you can't rerun time, fiddle with a few variables, and see what happens differently during our evolution. For us, our experimental laboratory is the past and the experiments have already been run. We are charged with making predictions backward—retrodictions—not about what will happen but about what has already happened that might be seen in the fossil or archeological record. We try to confirm or refute our hypotheses by inspecting and examining that record more intelligently. For experimental scientists, this may seem like a peculiar way of operating, but it is the primary method available to those who study the past.

Paleoanthropologists call any member of our lineage, including modern humans, a hominid or a hominin, depending on their views on taxonomy and naming of fossils. In this book, I am telling—retelling—the story of our lineage, and I don't want to confuse the matter with squabbles about technical details, so I will call any member of the genus *Homo* a human here.

The behavioral evolution of the hominid lineage can be boiled down to three big advances:

(1) toolmaking;
(2) language and symbolic behavior; and
(3) the domestication of other species.

If my hypothesis is correct, then at each of the three phases of human evolution demarcated by these advances, we'll be able to see clear evidence that our ancestors were vitally connected to animals. In other words, animals ought to be central to our way of life and our evolving behavior and anatomy. We ought to be able to see how relationships with animals benefited our ancestors and gave them a selective advantage—and we ought to be able to discern the growing importance of the links between animals and humans. Finally, we ought to be able to see how our connection with animals caused or contributed to these three big behavioral advances in human evolution. By the time you have finished reading this book, you should be able to judge for yourself if I am right or wrong.

To test my hypothesis against these predictions, I decided to follow the sage advice given by the King to the White Rabbit at the trial in *Alice's Adventures in Wonderland*: "Begin at the beginning and go on till you come to the end; then stop."

Using the White Rabbit principle, I'm going to start by looking at the very earliest members of the human lineage—the beginning—and then proceed chronologically until modern human times. At each stage, I'm going to discuss the changes that occurred and point out the role played by the animal connection.

The Animal Connection

1

Begin at the Beginning

THE BEGINNING of the hominid lineage—of species related to us—was roughly 6–7 million years ago in Africa. Before that time, there are no fossils anywhere in the world with the physical and anatomical traits that make them hominids, but later in time, there are plenty. Therefore, the lineage that led to modern chimps and bonobos and the one that led to us probably diverged or separated from each other sometime during that period.

I say "probably" because not every species that ever lived is preserved as a fossil. Anyone who studies evolution has to face the awkward reality that our fossil record will never be complete. Are divergence dates like these—ones based on the fossil record—reliable? If the fossil is older than about 50,000 years, radiometric dating can be very exact, because what is dated are the rocks above and below a particular fossil, not the fossil itself. And the chance that the very first hominid to walk the face of the earth would have been preserved and then found and then recognized is minuscule. So, can fossils really tell us when these two lineages split?

Fortunately, there is a second line of evidence that sheds light on the problem. By sequencing the DNA of different living species and comparing the DNA of one species to another, geneticists are able to count how many genetic mutations have occurred since the two species diverged in the evolutionary past. By counting the mutations in the same gene in different animals, geneticists access a sort of molecular clock in which each mutation represents a tick of the clock. Since different genes sometimes mutate at different rates—clicking faster or slower— this method works best when a number of different genes are studied and the results are combined. In the case of the chimp-human divergence, there is a great deal of molecular evidence which converges on a date between 4 and 6 million years ago, as the fossil evidence indicates. Two methods of estimating the divergence date for chimpanzees and humans concur.

The first species on the hominid line certainly wasn't a human, though, or anything much like you or me. In fact, exactly what the first hominid was like is still pretty fuzzy.

At present, there are three fossil species known from Africa, any one of which might be the earliest hominid: *Orrorin tugenensis*, which is dated to about 6 million years ago; *Sahelanthropus tchadensis,* dated by different techniques, which is believed to be older than 6 million years and perhaps between 6.8 and 7.2 million years old; and *Ardipithecus kadabba*, dated to between 5.2 and 5.8 million years ago. We don't yet have complete skeletons of any of these species, nor even a representative of each part of the body from different individuals, so I can't draw a detailed word picture of the adaptations of the earliest hominids. One, two, or all three of these early species might yet prove to be either early hominids or late common ancestors to both chimps and humans.

The two traits that let us classify these as early hominid candidates are that they have the right teeth and the right legs. All of these species have canine teeth that are more like those of modern humans than those of apes. Apes have larger, more daggerlike canine teeth than humans; our canine teeth are small and wear flat at the tip. Similarly, in all three species, bones from the lower limbs suggest that these species might have been bipedal, but we could use a lot more bones to be sure. Since our lineage includes the only mammals that habitually move about the world erect and on two legs that work in a striding fashion— not bouncing along with a hopping gait like a kangaroo—any bipedal mammal is a good candidate for hominid status. The way we walk is special and defining.

This conclusion is reinforced by another, more recent fossil (only 4.2 million years old) that is closely related to *Ardipithecus kadabba,* known as *Ardipithecus ramidus.* Fortunately, we have a lot of bones of *Ardipithecus ramidus*, including a partial skeleton of an adult female nicknamed Ardi. She was announced in the scientific and popular press with much fanfare late in 2009 and was named the discovery of the year by *Science* magazine. The analysis of Ardi and her kind provided a lot of surprises to the scientific community.

First, Ardi was unexpectedly large, almost as big as a male individual of her species. In later species like *Australopithecus afarensis*, best known from a female partial skeleton called Lucy, the females are much smaller than males, a condition known as sexual dimorphism.

Second, she had a very small skull relative to her body size and the small front teeth typical of hominids.

Third, she was found in a habitat that was forested, not open savannah (once thought to be essential to the evolution of bipedalism), and the other animals found with her were overwhelmingly forest animals, like colobus monkeys and forest antelopes.

Fourth, she walked upright on the ground in a completely novel fashion seen in no other known species, fossil or modern, yet probably clambered on all fours in the trees. If Ardi's type of bipedalism was typical of the other very old species of hominid, then the evolution of bipedalism took a very odd course indeed that no one would have predicted.

Finally, like many early hominids, *Ardipithecus* seems to have been a prey species. Many of the fossils, including those of Ardi herself, have been crunched and chewed by some predator, most probably the large catlike *Dinofelis,* which had impressive canine teeth and a large body size. What *Dinofelis* left behind after preying on the Ardi skeleton is hauntingly similar to what modern cheetahs leave behind when they eat a baboon: skull and teeth, arms complete to the hand, legs complete to the feet, and very little of the ribs, vertebrae, shoulder blades, or pelvis (Figure 1).

Ardi certainly could be called a bipedal ape with small teeth, except both bipedalism and small canine teeth make her a hominid, not an ape. This is a pretty unexciting description of the earliest member of the human lineage, but it is probably correct. Between those earliest hominids and the earliest humans lie a lot of evolutionary changes and several extinct species, some of which were on collateral lineages or side branches.

Deciding what is and isn't a new species in the fossil record can be very tricky. In biology, two populations are defined as separate species if they are reproductively isolated from one another: that is, if they can't interbreed and produce enough fertile offspring to survive in the long run. But fossils are admirably circumspect about their (former) sex lives. We have to depend, instead, on assessing the degree of physical difference between two fossil forms against the standard of the differences between living individuals of two separate species. We rarely find complete skeletons in the early part of the hominid record, and we aren't always lucky enough to find the same body parts so we can compare fossils to each other to see if they were adapted to similar lives and looked alike.

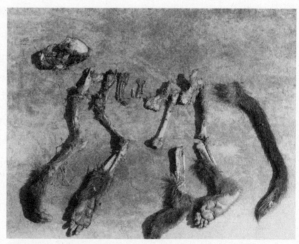

1. *The partial skeleton of* Ardipithecus ramidus *(left) is preserved much like the remains of a baboon eaten by cheetahs (above), suggesting that an extinct cat like* Dinofelis *was responsible for the missing portions of* Ardipithecus.

We have some excellent techniques for measurement, reconstruction, and comparison, but there is also a subjective element in assessing whether a new find is a new species. Paleontologists often joke that the basic rule is: If *I* find the fossil, it's a new species, but if *you* do, it isn't. The reality is that close, intensive study of a group of similar fossils tends to bias some researchers toward making many species because he or she sees a lot of individual variability in small details; such scientists are often called "splitters." How much difference in body weight, limb proportions, or tooth size—for example—are enough to make two generally similar fossils different species? Would a 20 percent difference in the grinding area of the cheek teeth be enough to separate two fossils into two species? Would 5 percent do—or 30? The answer is partly a judgment call.

Other researchers focus more on the features in common and the big picture of the adaptation as expressed in anatomy; these are called "lumpers." They are more inclined to look at differences in body size or tooth shape and compare those to the differences among living humans. So, for example, a lumper might say, "Yes, this fossil had arms that were about equal in length to its legs, whereas this other one had arms that were not as long as its legs. Since basically their anatomy says they were both long-armed species who swung through the trees like gibbons or siamangs, I will classify them as the same species, maybe in a single evolving lineage."

Let's try some concrete examples to illustrate this problem. If a jockey and a football player represent the extremes of human size and shape, and are clearly members of the same species, would two fossils have to show more difference than that to be two different species? Yes, if you have several complete, perfectly preserved skeletons of each size, which almost never happens. Even then, how many skeletons would you need to be sure? What are the chances that the fossils you happen to find are average-sized? What if by some statistical fluke the ones that get preserved are among the biggest, or the smallest, or both? These are thorny questions. This tendency leads to lots of disputes about terminology but relatively few about the visible anatomy and reconstructed behavior of the fossilized species in question.

Starting just over 4 million years ago, there are plenty of fossil hominids in Africa and none anywhere else. *Ardipithecus* was probably ancestral to some or all of the other hominid species in this period. Depending upon whether you are talking to a lumper or a splitter, there are at least seven and perhaps many more different species of hominid that lived in Africa between 4.2 million and about 2.5 million years ago. Sometimes two or three species lived simultaneously. Sometimes all of them went extinct but left descendants that became another species.

Most of these early hominids belong to the genus *Australopithecus* (which means "the southern apeman"). In roughly chronological order, the species in the genus *Australopithecus* include A. *anamensis*, A. *afarensis,* A. *garhi,* and A. *africanus*, though this is not necessarily a series of ancestors and descendants. A. *afarensis* includes the best-known individual in this group, Lucy, but there are thousands of less complete specimens of *Australopithecus* individuals. In casual conversation or popular writing, paleoanthropologists call the species in this genus "australopithecines" or, less felicitously, "australopiths."

At the same time, there were some quite similar species of early hominid with much larger, flatter cheek teeth (molars) and big crests for the attachment of chewing muscles on their skulls. Some paleoanthropologists put these fossils into the genus *Australopithecus*. In roughly chronological order these are: *A. aethiopicus, A. boisei,* and *A. robustus.* Others prefer the generic name *Paranthropus* for this group to emphasize the difference in body size and tooth size in the latter group; I will use the term *Paranthropus* here. Under either name, paranthropines and australopithecines were closely related to each other and very similar in many parts of their bodies. All were bipedal, though not in quite the way modern humans are, and they had brains as large or larger (relative to body size) than those of living apes. They were almost certainly hairy or furry, and their faces, their hands, and the overall shape and proportions of their bodies were not human. They had no clothes, no fire, no dwellings—and of course no domestic animals. And like apes, paranthropines and australopithecines seem to have been predominantly vegetarian (herbivorous in the broad sense), getting most of their nutrition from fruits, nuts, maybe tubers, and leaves or stems in varying proportions.

The heavier-built paranthropines differ from the australopithecines in having a very specialized anatomy related to their diet. The size of their cheek teeth (enormous), the thickness of their dental enamel (great), and the bony attachments for their chewing muscles (impressive) all suggest that they ate either tougher or harder foods than australopithecines. However, study of the microscopic wear on their teeth tells a more complex story. The food an animal eats wears the dental enamel, creating microwear. The patterns of markings—usually visible only under a microscope—are distinct enough that the dietary category (leaf-eater, meat-eater, hard nut–eater, and so on) into which an animal falls can be identified from the teeth alone. Despite the dental adaptations to eating hard (such as nuts) or tough (such as thick leaves and stems) foods, individuals whose teeth were studied showed microwear suggesting that these were fallback foods, consumed when enough soft foods, such as fruit, were unavailable. The South African paranthropines, in particular, apparently used hard nuts as a fallback food during only a part of the year.

Human ancestors or cousins they may have been, but neither australopithecines nor paranthropines were human. Some were in our lineage but they were not us.

Why not? Because they didn't have a big enough brain for their body size, they don't have cranial anatomy close enough to our own, and they don't seem to have exhibited behavior that was human enough—but where you draw that line is unclear.

Because toolmaking is the oldest behavioral trait of humans, and a fairly easy behavior to spot in the archeological record, I'd love to know which early hominids made and used tools.

This seemingly simple issue is terribly difficult to resolve. Simply because a species is found dead in the same place as tools doesn't mean that species was the toolmaker. As a somewhat frivolous example, let me explain that the animal most commonly found with early stone tools at sites in Olduvai Gorge in Tanzania is an antelope. If mere spatial association were enough to prove who was the toolmaker, then a lot of antelopes started making tools about 2.6 million years ago. Not very likely! And because more than one hominid lived in the same place at the same time that tools were being made, we have the additional problem of figuring out how many hominids made tools—and which tools were whose.

Just recently, a team of researchers led by Zeresenay Alem-seged, the curator of anthropology at the California Academy of Sciences, announced that they had found marks on two fossilized bones from 3.4 million years ago that they identify as cutmarks made by stone tools. This finding, if confirmed by additional work, is staggering. There are no stone tools known from anywhere in the world that are earlier than 2.6 million years ago, not even at Dikika, Ethiopia, where Alemseged's team found these bones. How do you get stone tool cutmarks without stone tools? That is the controversial question. Some people feel the team has misidentified the marks and they were made, perhaps by, crocodiles or carnivores. Others feel the dating of the bones is open to question, since they were found on the surface and not in excavation. Still others champion the team's finding, believing that this discovery indicates a very slow, opportunistic beginning in which sharp-edged rocks were occasionally used by early hominids who did not modify the stone further. Such tools are called "expediency tools" and would indicate tool *use*, not tool*making*. If so, then tool use was so infrequent for the first 800,000 years that it left very few recognizable indicators in the fossil and archeological record and did not have a visible effect on the hominids' biology until 2.6 million years ago. Only further work will resolve the controversy over this claim.

There is also no way either to prove or to rule out the possibility that *Australopithecus* or *Paranthropus* made stone tools before their extinction at about 1.5 million years ago, but I'll discuss some intriguing possibilities in the next chapter. It is true that, at just about the time that *Australopithecus garhi* and *Paranthropus aethiopicus*—which some paleoanthropologists think are the same species—first appeared in East Africa (2.6 million years ago), unquestionably recognizable stone tools also appeared. *Australopithecus garhi* is the same age as the earliest tools, which are accompanied by cut-marked animal bones, making it a good contender for the earliest toolmaker. Based on current evidence, the earliest *Homo*—*Homo habilis*—isn't as likely a candidate because it didn't appear in the fossil record until 300,000 years after the earliest stone tools.

Don't these temporal coincidences prove who made the tools? Well, no. Probably several species of hominid found the same sort of place attractive, so if tools were left behind by one temporary inhabitant, those tools might be found with a later inhabitant. Access to water or edible vegetation and places where the topography provided an extra measure of safety make for highly desirable location, as realtors say. Just as other mammals today use each other's trackways and dens, so hominids might have been drawn to the same locales. Kathy Schick, an archeologist at the Stone Age Institute, in Bloomington, Indiana, refers to this idea as the Favored Locale hypothesis.

Oddly, most paleoanthropologists would place the earliest stone tools in the hands of *Homo habilis* despite the fact that we don't yet have a *Homo habilis* fossil that is 2.6 million years old. There is an old and flawed expectation that only humans can make tools. In fact, *Homo habilis* was actually named "handy man" by Louis Leakey, Philip Tobias, and John Napier in 1965 because they believed that this then-new hominid was a toolmaker. They even included toolmaking in their description of the species. Since species are normally named and defined on a strictly anatomical basis, this action was considered reckless and certainly unwise at the time. But they may well have been right.

What made this ancestor of ours "human" (in the broad sense) when nothing earlier was? Some anatomical traits are key. Its face was more vertical than those of australopithecines and paranthropines, with smaller teeth: more like humans and less like apes. The vertical profile arose because the face of *Homo habilis* was tucked under its bigger braincase rather than being out in front of the braincase, as in apes or earlier hominids. That brain was also larger relative to body size than the brain of apes or of earlier hominids. The teeth were smaller than those in australopithecines and paranthropines and more humanlike in shape. Finally, the earliest members of the genus *Homo* walked upright and bipedally, with a pelvis much more like our own than the pelvises of other early hominids. Detail after detail of anatomy differentiates early *Homo* from *Australopithecus* or *Paranthropus*.

From the perspective of the whole course of human evolution, what really mattered about this ancestor of ours was something behavioral and not purely anatomical: the earliest members of the genus *Homo* almost certainly made stone tools, whether or not the other hominids did. Otherwise, when the australopithecines and paranthropines went extinct and only humans survived, the tools would vanish too. But they don't. Humans and tools continue unabated. Thus, toolmaking is strongly tied to humans by some of the chronological evidence as well as by human prejudice and pride in being a toolmaker.

Tools changed hominids a great deal—or at least, they changed the ones we call humans, the ones ancestral to us.

Strictly speaking, the invention of stone tools should be identified as the signature of something that might be—but generally isn't—called the Paleolithic Revolution. After humans learned to make stone tools, things were never the same again for our lineage.

2

Evolve Without Evolving

LEARNING HOW TO make stone tools was a first huge step for the hominid lineage and an extraordinary intellectual breakthrough. It represented much more than learning a set of manual skills with which to shape raw material. What this invention signaled was the realization, first, that the apparent attributes of an object are not all there is; and, second, that humans had the power to transform those attributes. Tools enabled our lineage to evolve—to change our behavior and ecological niche—without evolving physically.

What do I mean by this?

Pick up a stone, any stone, about the size of a small loaf of bread. What do you notice? The stone may be longer or shorter, thinner or thicker, or rounder or more angular depending upon the type you have picked up. Let's suppose you have chosen a good raw material for making a stone tool, something like a handy river cobble. The apparent properties of this object in your hands are its color, its general size, and its heaviness. If you have an analytical turn of mind, you might think it is dense: heavy for its size. If you have picked up a river cobble, the object in your hands is probably blunt and more or less rounded on the edges.

If you are a toolmaker, however, you also know that concealed inside this blunt, heavy, large object is a lighter, smaller object with sharp edges. A toolmaker can recognize a stone that can be transformed into an object with radically different properties if it is hit with another, smaller, handheld rock in a particular manner. And by carrying out a sequence of careful blows, you as toolmaker can change the apparent properties of the cobble in a reasonably short period of time.

None of these facts is obvious.

Realizing that you can transform the properties of the stone through your own actions is little short of brilliant. What this means is that what you see is not what you get—you can get something else. Since rocks and stones appear so strong, so unyielding, so unchanging, figuring out that they can be changed is a much more fundamental surprise than figuring out that a nut with a hard shell can be broken to extract the meat within. Nuts break naturally and are opened daily by other animals, like squirrels or birds. Broken nuts are still nuts—reproductive parts of trees protected by a shell or outer covering; they don't turn into something entirely different when they are broken, though the inner part of the nut is usually softer and much tastier than the shell. Although natural forces can and do break rocks, we don't observe that happening every day.

Once you understand that a rock can be transformed, you still don't know how to make a tool. Learning how to effect these transformations is not simple and was probably harder for the first hominid on the face of the earth to try this task. Even if someone who already knows how to shape rocks shows you, transforming a rock into a tool is not trivial. I've tried it. Believe me, the process of deliberately shaping rocks is not easily mastered. The most primitive technique for doing this is known as knapping, chipping, or flaking, and it is difficult. There are a lot of sore fingers, scraped knuckles, aching arms, and failures before you reach the point where enough practice might produce mastery.

First you have to recognize the right sort of stone (raw material). Different raw materials react very differently to knapping. The most important criteria in choosing a rock from which to make a stone tool are its grain size and crystallinity. For example, fine-grained lavas make good tools, but coarse-grained ones do not. One of the most difficult raw materials from which chipped stone tools have been made is quartz. Quartz grows naturally in crystals. The flat planes of the crystals are also inherent planes of weakness that will separate from the next crystal when you hit the quartz with another rock. Unfortunately, those planes of weakness won't be where you want the rock to break. At the other end of the spectrum are rocks like obsidian, which is a black volcanic glass with an amorphous internal structure and no crystals. There are no natural planes of cleavage in obsidian, so it can be made into exquisitely sharp tools. In fact, obsidian blades are sharper—have a narrower blade—than steel scalpels and have been used in cardiac surgery. Flint and chert—other great choices for making flaked tools—are cryptocrystalline, meaning that their crystals are so small they are difficult to see even microscopically. When rocks like flint or chert are hit, correctly, with a hammerstone, the impact of the hammerstone produces a cone of force that travels through the target rock, without being deflected by crystals, and produces a curving fracture. Such fractures are called conchoidal because they somewhat resemble the curving shape of a shell.

I don't think the first hominid to knap stones recognized that some were crystalline and other weren't. I doubt he or she was thinking about fracture planes or conchoidal fractures or grain size or angles at which one stone might strike another. I find it impossible to believe that the first human toolmaker had words or concepts for flint, obsidian, chert, quartz, not to mention conchoidal. But what that individual did recognize was what worked. Somehow early humans began to realize—learn—which kinds of stones flake well and which poorly, which ways of smacking one stone into another produce the desired trans-formation of the properties of the core and which are useless (Figure 2).

What you choose as a raw material for the target rock or core is important to your success, but so is the material you choose to use as a hammerstone—the handheld rock with which you strike the core. The hammerstone needs to be the right size for easy use, roughly ovoid (to fit into the hand) but not sharp-edged, and heavy enough to help you produce an effective blow.

2. *A right-handed person makes a stone tool by striking the core (held in the left hand) with a hammerstone (held in the right hand), as shown by the black arrow. As each successive flake is removed, the core is rotated in a clockwise direction (white arrows) to remove new flakes. Each flake has a curved surface known as the bulb of percussion, as well as remnants of the striking platform (shown in stipple).*

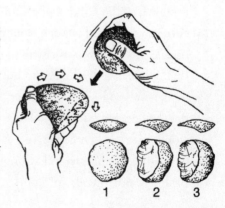

1   2   3

Quartzite, a hard sandstone, and limestone are common raw materials for hammerstones. Because the hammerstone is hard, this technique of making tools is termed *hard hammer percussion* to distinguish it from a more advanced technique (*soft hammer percussion*) in which objects like bones or antlers are used as hammers.

Once you have found and assembled the right raw materials in a convenient spot, you still have to learn the technique and manual skills. Basic procedures were used to produce the earliest primitive stone tools, a type known as the Oldowan industry, after Olduvai Gorge, Tanzania, where such tools were first recognized by Mary Leakey. Mary had no formal training in archeology, but an artist's eye and a scientist's acute brain led her to become recognized worldwide as a leader in archeology. She was one of the best archeologists and smartest women I ever met.

The process of making a tool sounds fairly simple: you strike the core with a hammerstone, aiming at a flat or roughly flat surface, and bingo! off comes a nice sharp flake. Only that isn't exactly true.

The place where you strike the core with the hammerstone is known as the striking platform. If you hit the striking platform correctly and hard enough, you will remove a flake of stone. The trick is that you want a striking platform that is angled at less than 90 degrees to the rest of the piece and your eye has to learn to detect good striking platforms. If a good striking platform doesn't occur naturally, you have to prepare one by breaking the stone just right. Once you've got a good platform, if you repeatedly strike it sharply and strongly with the hammerstone—not bashing your fingers, of course—then each blow will fracture the core conchoidally, removing another flake, until you run out of platform.

Flakes are the razor blades of stone tools: small, light, easy to maneuver, and, if your raw material is well selected, incredibly sharp and as likely to cut your own fingers as the substance you intend to cut. They are not much like the rock you started with. Flakes can be recognized by traces of the process that produced them: the remnant of the striking platform and the characteristic swelling on the underside of the flake, where it has separated from the core. This swelling is known as a bulb of percussion.

The other product of repeatedly striking the core may be a core tool—what is left of the original core after all the flakes have been removed. In the Oldowan, sometimes these cores were used in power tasks that required heavy-duty tools, such as chopping or hacking apart the joints of large dead animals. Other times, the core seems to have been a waste product: the package that once held a lot of razor blades.

From beginning to end, the process of making a stone tool is one of reduction. You reduce the bulk of the core to leave only pieces suitable for your task. But the conceptual leap and the physical learning behind this "simple" process were anything but simple.

A tool is defined as "a contrivance or device for doing work." The word "tool" itself is derived from the Old English *tawian*, meaning "to prepare" or "make." Paleoanthropologists make an important distinction between an object that is used briefly without modification or with minimal modification—an expediency tool, like those suggested to have been used at Dikika 3.4 million years ago—and a tool which has been deliberately shaped or otherwise modified in order to better perform its intended use. The latter are termed *artifacts*, not to be confused with the word "artifact" to mean an inaccurate observation, effect, or result.

Thus, if you take a branch that has been blown down by the wind, trim it to the desired length, and sharpen its point to use as a plant stake, that is a tool in the archeological sense. Someone coming along later will be able to see the modifications (trimming, sharpening) you have made and will be able to recognize that this shaped branch was a tool. Those modifications are physical evidence of your intentionality and planning in the same way that the extensive flaking on the Olduvai stone tools attests to a hominid's foresight and planning. On the other hand, if you pick up a rock to whack this newly made plant stake into the ground, the rock is an expediency tool—a used but unmodified object—and would be very difficult to recognize later. But if that rock was used repeatedly to whack plant stakes into the ground, in time the use wear would distinguish the rock as utilized, even if it had little shaping beforehand.

Tools are objects outside of (external to) the tool user that are used for a purpose. In other words, tools are extrasomatic adaptations: ways in which a species adapts to its world without evolving. This is a key concept for understanding human evolution.

The possible gain from an extrasomatic adaptation is huge. Potentially, change can occur much more rapidly through extrasomatic adaptations than through the normal processes of selection and evolution.

For example, suppose a nutritious resource—such as an oily, carbohydrate-rich nut inside a hard shell—became more common in the area in which an animal lived. One way of taking advantage of this new and important resource would be to evolve larger jaw muscles that could be used for chomping down on and opening this nut. Another might be to evolve changes in your forearm and hand anatomy that would let you successfully break into these nuts with your bare hands. Such changes might take many, many generations to evolve under natural selection. In contrast, a tool-maker's way of doing the same thing might be to pick up a heavy branch or rock and use it to smash open the nut. Understanding the concept of using a tool to do something that can't be done without a tool is an enormous intellectual breakthrough. Once the concept of tool using is achieved, the gain is immediate.

Extrasomatic adaptations free tool users from the tyranny of biological evolution to a large extent. In terms an ad agency might use, the concept is: "Is the climate growing colder? Don't evolve thicker fur. Learn to wear the fur of others!" The same kind of shortcut adaptation is easily envisioned as a solution to myriad needs. Fire is another kind of tool for keeping warm— and for processing food to make it more edible. Cutting and shaping wood or meat and hide can be more easily achieved by making stone tools than by evolving sharper teeth with larger cutting blades on them. Containers and baskets or hide sacks are another major advance that transforms the task of gathering and transporting many small objects (such as shellfish, berries, nuts) or liquids (water, milk) into something much simpler. You may have seen comical images of some monkeys stuffing their cheek pouches with fruit, which is their evolutionary answer to carrying small food items. But a simple sack or even some skill-fully folded leaves work as well or better for transporting many small fruits than cheek pouches and doesn't take generations to evolve.

3. *This Oldowan chopper is a core tool made by removing flakes. This specimen comes from Gona, Ethiopia—the oldest known stone tool site in the world—and is 2.6 million years old.*

Extrasomatic adaptations can have dramatic effects. Inventing stone tools was the first great transformation of the hominid lineage and that invention changed our lineage—and us—irrevocably. And starting to adapt extrasomatically gave us a mechanism that could change the habitat, the environment, even the world around us. Our ancestors embarked on creating their own ecological niche, a behavior known as "niche construction." To see how this was done and why it paid off, you need to look at how stone tools were made and what they were used for.

My friend Sileshi Semaw of the Stone Age Institute leads the team that discovered the oldest known archeological sites in Gona, Ethiopia, which date to 2.6 million years ago (Figure 3). Ethiopia is Sileshi's home country and one of the richest regions of the world in terms of the early paleoanthropological record.

Sileshi's group has found more than a dozen archeological localities in the Gona region that have yielded early stone tools.

The groups of stone tools from each of these sites—what an archeologist would term the *lithic assemblages*—are rather consistent. There are always cores (the basic piece of stone from which flakes have been removed), some of which may have been used as hammerstones. There are a large number of flakes (both whole and broken) and many angular fragments. Sometimes there are manuports—rocks carried to the site from elsewhere—that may have been intended for use as cores. Only a small number of the deliberately flaked tools were retouched or resharpened after the initial flaking.

The objective of Oldowan tool production was the manufacturing of sharp flakes, says Sileshi, and most cores were flaked only on one side. Sileshi sees competence and mastery of flake production in all of the assemblages earlier than about 1.5 million years ago. He thinks the variability within the stone tools is due to the different raw materials at hand and their different flaking properties.

From his perspective, the Oldowan lasted for over 1 million years with little or no change. Sileshi argues that this is simply technological stasis. Oldowan people were competent and knew what they were doing, but after the initial breakthrough, they don't seem to have been very innovative.

All of this information about making stone tools is interesting, but it doesn't touch on the crucial issues. Why did hominids want small sharp flakes? What are stone tools *for*?

Specimens from Gona and at sites at Bouri, also in the Awash River Valley of Ethiopia, give the answer. Seven individual sites from Gona that have yielded artifacts were also found to contain the bones of several large animals. Although the bones are few in number, four of the Gona sites include at least one cut-marked bone, as well as the tools that probably caused the mark. Other Gona sites contain artifacts only, probably because local conditions were not good for bone preservation everywhere. I have often said, jokingly, that stone tools never die, but bones do. Stone tools last forever, while bones are often eaten, decayed, trampled, or destroyed by natural processes.

Bouri is a nearby 2.5 million-year-old site discovered by a team led by another friend, Tim White, of the University of California, Berkeley. Tim's team recovered more than four hundred fossilized bones of animals, including ones that were broken and cut-marked. However, at Bouri the cut-marked bones were not associated with dense concentrations of stone tools. Flakes and cores occurred as isolated and widely scattered specimens, even though those flakes or similar ones must have been used to produce the cutmarks.

The teams at Gona and Bouri knew the marks they saw on the fossilized bones were cutmarks because of work I and others did in the early 1980s. I was one of the first paleoanthropologists to show that cutmarks could be distinguished from animal toothmarks and marks produced by digestion, weathering, abrasion, or trampling on the basis of their microscopic characteristics. The research project began in the late 1970s, when Rick Potts, now at the Smithsonian Institution, and I were both in Nairobi working on the Olduvai collections, sitting across from each other at the same table.

One of us noticed some funny marks on the femur (thigh bone) of a massive fossil antelope and passed the bone over to the other.

"What do you think those are?" one of us said.

"Cutmarks!" came the reply.

We both intuitively knew what we were seeing, but we didn't know how to prove it to a skeptic. So we set about figuring out how cutmarks differed from other similar kinds of marks that might appear on the bones of fossil species.

The key was doing a lot of experiments and looking at the resulting marks under magnification (Figure 4). Though many events might produce a groove on a bone, only stone tools consistently make a V-shaped groove that has fine, parallel striations running down the sides and bottom of the groove. The linear striations are drag marks or indentations made by each tiny point or irregularity on the sharp edge of the tool. In contrast, animal toothmarks are generally smooth-bottomed because animal teeth are generally smooth on the outside. The exception occurs when repeated gnawing or chewing creates marks over marks, but those cases are usually obvious. Trampling on a bone

4. (*top*) An experimental cutmark is a V-shaped groove with fine, parallel, linear striations. (*middle*) A carnivore's toothmark is often U-shaped and lacks internal striations. (*bottom*) A mark of unknown origin on a fossilized bone from Olduvai Gorge, Tanzania, shows the same features as the experimentally made cutmark and is judged to be a cutmark.

lying on rocky ground does produce linear striations, but these also tend to occur all over the entire surface of a bone, not in just a single area. Trampling doesn't leave a largely pristine bone surface with only a few marks as cutting or scraping does. We tried many other possibilities, but cutmarks kept coming out as distinctive—and kept matching that mark on the antelope femur from Olduvai. After a lot of experimentation, we felt we had proof positive that hominids used stone tools to process animal carcasses. Today, our techniques are widely accepted, though initially they were controversial because our critics didn't think we'd looked at enough potential mimics of cutmarks.

In subsequent years, Rob Blumenschine of Rutgers University and some of his students at the time—Marie Selvagio, Sal Capaldo, and others—followed our lead, finding diagnostic characteristics of percussion marks made on bones with hammerstones. Additional studies carried out by Travis Rayne Pickering, now at the University of Wisconsin at Madison, and Charles Egeland, of the University of North Carolina–Greens-

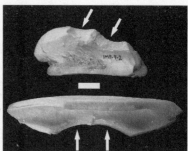

5. *Using percussion to break open a marrow bone, by striking it with a hammerstone, leaves two types of marks: (left) microscopic distinctive percussion marks; and (below) beveled notches.*

boro, refined the identification of percussion marks further. Hitting an animal's long bone with a hammerstone may break it, leaving a diagnostic negative flake scar on the inside (the marrow cavity) of the bone and anvil scratches on the bone's underside. Even if the bone breaks into many pieces, you can put the fragments back together and see the negative flake scar (Figure 5). The point of breaking most bones is to get to the highly nutritious and fatty marrow that is inside the long bones of most animals. This experimental and comparative approach to understanding modifications and damage on bones has now become standard in analyzing fossil and recent bones.

As other workers followed similar lines of research over the last few decades, some marks have been documented that can be hauntingly similar to cutmarks. One of the agents that leave marks requiring the most careful interpretation are crocodiles, studied by Jackson Njau, now a postdoctoral fellow at the University of California at Berkeley. Since crocodiles grab their prey and spin, twist, and wrestle it in order first to drown and then to dismember it for handy eating, crocodiles produce a wide variety of marks. Too, when crocodiles damage their teeth in these highly violent attacks, the resultant marks take a wide range of forms that can resemble cutmarks. Interpreting marks on bones—once fairly simple—has become much more complex as our knowledge of the natural processes that damage bones becomes more extensive.

The cut-marked, ancient bones from Gona and Bouri show that the toolmakers used their new, sharp implements on dead animals, probably to remove flesh, hide, and fat. Paleoanthropologists have found cutmarks on many fossil bones ever since Rick and I first noticed the ones on that femur from Olduvai—

which, to be fair, Mary Leakey had recognized before us, though she didn't undertake the comparative work we did or a systematic survey of Olduvai bones looking for cutmarks as I did.

Finding such marks at Gona and Bouri is particularly important because these are the oldest hominid sites with stone tools. The marks are evidence of their use in the world. And all of the cutmarks and percussion marks left by these stone tools occur on animal bones. From the very beginning, there is strong evidence that flakes were used in processing—cutting up—dead animals. If the Dikika marks are correctly identified and dated, then they attest to this use of sharp stones even earlier. Cutmark evidence becomes abundant about 2.5 million years ago and continues until the relatively recent past, when marks made by metal knives and implements replace those made by stone.

Couldn't early tools also have been used to obtain vegetable foods or to cut wood? Yes, they might have. But the evidence that early tools were used on plant materials is extremely scanty compared to the evidence that they were used on dead animals. Only two studies and eight individual tools provide evidence of early tool use on plants.

The logic underlying those studies is similar to the logic behind identifying cutmarks through microscopic features. When a stone tool is used, the substance upon which it is used leaves either residues or modifications (damage) on the tool. In forensic anthropology, this idea is known as Locard's exchange principle: every contact between two objects or two people involves an exchange of traces. Diagnosing exactly which trace or damage was caused by what action is extremely difficult and requires a large comparative sample. Many uses produce similar damage; residues can be altered or obliterated by various natural processes; and some raw materials (like lava) simply do not hold detectable microwear. Out of fifty-four Oldowan artifacts from Koobi Fora in Kenya, only nine had microwear on their working edges. Four were diagnosed as having been used in meat-slicing, two were used on reeds or soft plant materials, and three were used on wood. In a similar study, five handaxes—archetypal tools of the Acheulian industry that followed the Oldowan—from Peninj in Tanzania were examined for microscopic traces of plants. Three were found to have phytoliths (microscopic silica particles from inside plant cells) on their working edges and may have been used on *Acacia* wood.

Since both the Koobi Fora and the Peninj tools are dated to about 1.5 million years ago, they don't necessarily prove that plant materials were regularly processed with tools from the very beginning. They do show that later stone tools were used to process plant remains.

In establishing the function and use of early stone tools, many paleoanthropologists have turned to the Olduvai collections because the fossils and stone tools are numerous and beautifully preserved. The thousands of specimens from Olduvai enabled scientists to survey numerous bones and stone tools in order to use statistical tests to explore the associations among the bones and stones.

I was able to show that the number of stone flakes in each of the sites in Bed I, Olduvai—the oldest layer—is statistically correlated with the frequency of cut-marked bones at that site. More flakes equals more cutmarks. This seems obvious, but it is just another piece of information reinforcing the idea that the primary function of stone tools was to process dead animals.

I also found that the Olduvai cutmarks aren't concentrated

on animals in a restricted size range. Cutmarks and percussion marks have been found on animals at Olduvai as small as hedge-hogs and as large as elephants. Prey size at Olduvai, as judged from the presence of cutmarks and percussion marks, covers what was probably the full range of everything that lived at Olduvai, from very small mammals (hedgehog) through small-, medium-, and large-sized antelope (gazelle to buffalo), to very large animals (giraffe or elephant). The abundance of cutmarks on the different prey species mirrors the proportion of those animals in the fossil assemblage. In other words, the hominids were not targeting a particular species or a particular size of species.

That fact is remarkable.

Carnivores focus their hunting on a restricted size range of animals, predictably based on their own body size. Norman Owen-Smith of the University of Witwatersrand and M. G. L. Mills of the University of Pretoria carried out a massive analysis of predation at Kruger National Park in South Africa from 1954 to 1985 involving almost 48,000 kills. Kruger is home to five main carnivores—lions, spotted hyenas, leopards, cheetahs, and Cape hunting dogs—and twenty-two species of herbivores larger than 22 lbs (10 kg) as adults. The Kruger carnivores concentrate on prey species that range from about half their own body size to twice as big. Carnivores that hunt in social groups tend to take slightly larger prey than would be predicted for a solitary species.

The five Kruger carnivores studied by Owen-Smith and Mills group into three medium-sized species (Cape hunting dog, cheetah, and leopard) and two large ones (spotted hyena and lion). The Cape hunting dog weighs only about 60 lbs (27 kg); the cheetah is roughly twice as big; and the leopard is a bit bigger still (about 135 lbs or 61 kg). As the smallest of these carnivores, you'd expect the Cape hunting dog to take the smallest prey, less than 22 lbs (10 kg) or so. But because the dogs hunt in packs, they take bigger prey than would be expected. Fully 90 percent of the wild dogs' prey is in the small antelope size class. The cheetah and leopard also focus on prey of small antelope size, even though the cats are bigger-bodied, because they are mostly solitary hunters. Eighty-five to 90 percent of prey taken by these cats are impala or other small antelope weighing roughly 45–90 lbs (20–40 kg). There is a lot of competition for those small antelope!

The two largest carnivores in Kruger are lions (275–600 lbs or 126–272 kg) and spotted hyenas (100–175 lbs or 45–80 kg). Both are social hunters and both take species larger than themselves. About 80 percent of the animals killed by spotted hyenas are the size of small antelopes and 15 percent are the size of medium antelopes. Lions' prey is 50 percent small antelope–sized, 27 percent medium antelope–sized, 15 percent big antelope–sized, and 10 percent very large antelope–sized animals. In other areas of Africa, lion preferentially prey on buffalo and giraffe, which are even larger. Lions are the only predators in Kruger to regularly take down megaherbivores like giraffe, hippo, or elephant (1,800–6,000 lbs or 825–2,800 kg) (Figure 6).

Data like these from real-life observations were used to derive equations that predict the size of prey a carnivore will take based on its own body size. Additional studies by Chidi Nwokeji, a student at the University of Frankfurt, have led to equations in which both body size and whether a species hunts singly or in groups can be factored into predictions of prey focal size. Using these equations, I can estimate what size of prey early hominids would be expected to take if they behaved like

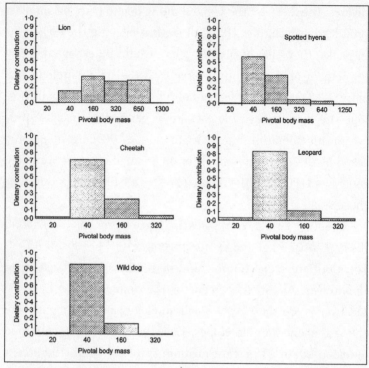

6. *Modern predators in Kruger National Park include large-sized (lion, spotted hyena), medium-sized (cheetah and leopard), and small-sized (wild dogs) judged by body size. The histograms show the portion of the predators' diet made up of prey species in different-size classes (left to right, small to large). Larger predators focus on absolutely and relatively larger prey. Small predators take a narrower range of smaller prey.*

carnivores. Even though paranthropines, australopithecines, and early humans were not true carnivores, following this procedure gave me a good idea of how early hominids stacked up to real carnivores.

Paranthropines, australopithecines, and the earliest humans weighed about as much as the Cape hunting dog or the cheetah, and a bit less than most leopards: roughly 60–108 lbs

(30–49 kg). Let's assume for the moment that the stone tools were wielded by early *Homo*, though all hominids alive at the time in East Africa were fairly similar in size. If humans were hunting singly, then they should have focused on prey in the 44–88 lb (20–40 kg) class, like *Antidorcas recki*, a relative of the living springbok that was found in Olduvai 1.5 million years ago. If humans hunted in groups averaging five individuals, for example, then their focal prey size should have been about the size of a 100 lb medium antelope, such as nyala or the extinct waterbuck at Olduvai.

The cutmarks attest to something extraordinarily different from these expectations. Let's focus on a single site from Olduvai, the one that has the most compelling evidence for hominid involvement with the fossilized bones: FLK Zinj. At FLK Zinj, there are stone tools and thousands of bones, as well as bones of two hominids: *Homo habilis* and *Paranthropus boisei*. The animal bones show lots of cutmarks and percussion marks. Because many of the animal bones are broken into pieces, but can be fitted back together, it is very probable that the animals were brought to the site and processed there by tool-using early hominids.

On the very small end of the size scale, there are about 16,000 bones of micromammals—many individual rodents and the like—none of which bear cutmarks. Cutmarks are generally very rare on micromammals anyway because you don't need to cut a mouse up to eat it. (Ask any cat.) At the top end of the size scale, FLK Zinj preserves bones of giraffe, hippo, and elephant, none of which have cutmarks. We know that neither the very small species nor the very large species were outside the realm of possibility, because bones of both from other Olduvai sites do bear cutmarks.

7. *Of all the antelopes at the FLK Zinj site at Olduvai (at least twenty-nine animals, shown by shaded bars), at least eleven individuals show cutmarks (open bars). The approximate body size of each species is indicated by the color of the silhouette; small antelopes (88–350 lbs) have open silhouettes; medium antelopes (350–700 lbs) are in light gray; large antelopes (700 lbs–1,430 lbs) are in dark gray; very large antelopes (1,430 lbs or greater) in black. MNI = the Minimum Number of Individuals from which the bones came.*

The preferred prey for any carnivore living in FLK Zinj times would certainly have been some kind of antelope. There are at least twenty-nine individual antelopes represented at FLK Zinj, ranging in size from the dainty *Antidorcas* to the hefty *Syncerus* (a buffalo of about 1,000 lbs or 480 kg) (Figure 7). These antelope bones have been scrutinized for cutmarks and percussion marks by several paleoanthropologists, myself included. Of those bones bearing cutmarks at FLK Zinj, bones of small antelopes comprise 24 percent, medium antelope bones make up 31 percent and a startling 41 percent of the bones are for large antelopes up to about 700 lbs. A final 3 percent of the cut-marked animals belong to the very large extinct buffalo or similar-sized antelopes.

Compare this distribution of prey sizes to those taken by a cheetah—with a comparable body weight to that of the hominids—and you can see how strangely early humans behaved. A cheetah takes about 5 percent very small animals, 85 percent small antelope, 7 percent medium-sized antelope, and maybe 2 percent of large antelopes. Instead of taking a range of prey similar to a cheetah, the humans using tools at FLK Zinj hunted more like lions than any other still-living predator. However, a lion weighs from four to ten times as much as an early human.

How could humans have escaped the rules of predator-prey behavior that are mathematically predictable among true carnivores?

Why did humans hunt like animals that were so much bigger than their own body size?

For that matter, how could humans or other hominids have functioned as carnivores at all with their feeble anatomy?

Feeble? Yes, our ancestors were less than physically impressive. They lacked the adaptations of speed, strength, sharp claws, slicing teeth, acute hearing, and keen sense of smell that make carnivores so effective. Astonishingly, early humans had only two apparent physical adaptations to hunting: an ability to throw with reasonable accuracy, and stone tools. These were the vital adaptations that enabled humans to hunt so much more effectively than other animals their size.

You might wonder if early humans' unexpected skill as hunters came from their primate brains. Primates are notoriously big-brained and clever animals (or so we like to think). Though we don't know much about the last common ancestor between chimps and humans, we can look at the hunting habits of modern chimpanzees as a way of making sense of the archeological record. Chimps are our closest living relatives, are only slightly larger in body size than early humans, and—like early humans—do not have strong physical adaptations to hunting. Remember, though, that both the hominid and chimpanzee lineages have evolved since they diverged from each other 6–7 million years ago. Chimps may well be as different from their earliest ancestors as we are from ours.

Chimps are absolutely hopeless carnivores. Though hunting among chimps has received a lot of attention, it is really a very minor behavior. The major part of their diet is comprised of ripe fruits and nuts; chimps spend about 60 percent of their time feeding on fruits, another 22 percent eating nuts, about 5 percent eating seeds and blossoms. Seasonally, females and young may fish for termites or ants, but males rarely do, and other insects are eaten incidentally. Chimps also eat small lizards, baby monkeys, newborn gazelles, and eggs as minor dietary components.

Only about 5 percent of a chimpanzee's annual diet comes from hunting, probably because the number of kills is so small. We know that the activity of hunting occupies only about 1.5 percent of chimps' total feeding time. Because they hunt in groups and weigh 60–150 lbs (26–70 kg), we can say that if chimps were carnivores, they should focus on animals of 44–88 lbs (20–40 kg) in body size. Instead, the most common prey of chimps is the juvenile red colobus monkey, which weighs a paltry 22 lbs (10 kg) or less. In other words, chimps are such poor hunters that they take prey that is only half as big as would be predicted if they were true carnivores.

This ineptness is not just a chimp problem. A similar limitation is seen among baboons, the common large-bodied monkeys of Africa. Though olive baboons (*Papio anubis*) weigh about as much as early humans, and hunt in groups, their focal prey is even smaller than that of chimps: only about 4–10 lbs (2–3.5 kg).

Archeologist Tom Plummer of Queens College, City University of New York, and primatologist Craig Stanford of the University of Southern California believe chimp hunting is *directly* limited by their lack of tools. "Chimpanzees are limited to prey items that they can pursue, capture, kill, disarticulate and consume without technological assistance," they say. Chimps lack the bodily adaptations to be effective carnivores, and because chimps also lack the right type of tools, they are terrible hunters. They are not able to gain enough food from hunting to make much of a difference to their diet.

Tools make all the difference.

Tool-using hominids hunted not like chimps or baboons and not even like ordinary carnivores. Tool-using hominids hunted like supercarnivores.

Tools enabled hominids to step outside the normal rules of predators and prey so they could function like a much larger and much more able carnivore. By inventing tools, hominids circumvented the long evolutionary road that made lions lions or hyenas hyenas. Hominids were able to take a shortcut from being a large-bodied primate that occasionally captured a small animal to being a serious competitor for large prey. This is the kind of huge effect produced by extrasomatic adaptations.

But was the emergence of hunting skill really so abrupt, or were hominids already functioning as hunters before the invention of stone tools? Couldn't early hominids and humans have made tools out of wood or plant material?

3

Attention Must Be Paid

WE DON'T KNOW how long it took our ancestors to become adept hunters. What we do know is that the concrete evidence of cutmarks and percussion marks proves that early hominids exploited a surprisingly broad range of prey as soon as they knew how to make tools. The comparative evidence of the hunting abilities of chimps or baboons strongly suggests that before the invention of stone tools, hominids were not successful or regular hunters.

Can we tell how our ancestors obtained their prey? Both scavenging and hunting have been suggested as early hominid lifestyles. No living mammals are exclusively scavengers, so a reasonable expectation would be that hominids hunted some and scavenged some.

One way of figuring out how hominids got their meat is to look at the distribution of cutmarks on the bones of their prey. Cutmarks on Olduvai antelopes from the FLK Zinj site are not distributed randomly at all (Figure 8). They occur preferentially on the forelegs, particularly near the elbow (on the humerus, radius, and ulna). If meat were the sole objective, then most of the cutmarks ought to occur on the meatiest portions of an animal: the humerus of the foreleg and the femur of the hindleg. Indeed, a large proportion of cutmarks (62 percent on the small antelope and 39 percent on the medium and large antelope) do occur on those meat-bearing bones. But bones with little meat are also often cut-marked, suggesting that hide or tendon was being removed as well as meat.

Did hominids simply take the leftovers after the real carnivores were done with a carcass? Carnivores usually eat the meatiest parts of the forelimbs and hindlimbs of kills first, which means there isn't much left for scavengers on those bones. Hungry hyenas leave very little after they have finished with a carcass, because their powerful jaws can slice meat and crack bones. Lions, leopards, cheetahs, and Cape hunting dogs leave more behind, because (except with the smallest prey)

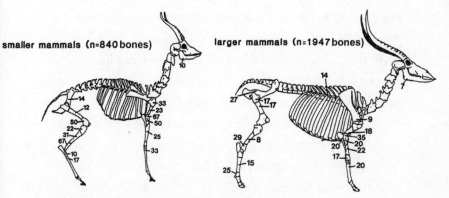

smaller mammals (n=840 bones)   larger mammals (n=1947 bones)

8. *Early hominids more often left cutmarks on particular joints than on others. The numbers next to the skeleton drawings indicate the percentage of cutmarks found at that point on antelopes from the FLK Zinj site at Olduvai Gorge. Note that the elbow and knee (stifle) joints are often cut-marked.*

their jaws aren't strong enough to crack bones. Still, only small pieces, scraps and shreds of meat or hide are left when these carnivores have finished with a kill.

Because we find many cutmarks on the meaty parts of the forelimb, we know that the hominids had access to carcasses before the carnivores had eaten everything. Possibly the hominids were killing the animals themselves, which would account for these facts. Alternatively, if they were scavenging carcasses rather than killing them, then early hominids were able to steal meaty portions from the carnivore(s) that killed the animal before all the meat was eaten. This behavior has been called power scavenging by Henry Bunn of the University of Wisconsin.

Another point confirms that direct competition over carcasses was common at these early sites. Many bones bear both carnivore toothmarks and cutmarks. On some individual specimens, cutmarks and carnivore toothmarks actual overlap each other. Sadly, there are so few of these overlapping cases that you can't figure out a consistent pattern of either carnivores-first or hominids-first. But there certainly was direct competition going on.

At FLK Zinj, antelope bones that were part of a limb segment from a single animal were statistically more likely to have cutmarks than isolated bones. This pattern suggests that hominids hacked off a portion of a limb and carried it away from the carcass, rather than staying and continuing to fight off other predators and scavengers. Because close observation of other animals was advantageous as soon as stone tools were invented, hominids modified their carcass-processing behaviors to lessen their exposure to competition from other carnivores. We can actually see evidence of a hack-and-run behavior in the fossilized animal bones!

Percussion marks on limb bones also show that hominids were breaking bones open and extracting marrow. Marrow is fatty and rich, and today hyenas are usually the only African carnivore to be able to break open the bones of medium- to large-sized antelopes. However, a hominid carrying a stone tool can crack the bones easily and extract a wonderful nutritional resource, even after all the meat from a carcass is long gone. Hominids may have been competing with carnivores for the meat, but they were also leaving signs that they could use parts of the prey (marrow, hide, or tendon) that carnivores often discarded.

Though there is much that we don't know, we can conclude something hugely important from all of this evidence about stones and the cutmarks and percussion marks that link the stones to the bones. Tools gave our ancestors access to new sources of food that were rich in proteins and fats: nutrients not found in large amounts in many vegetable foods. What's more, animal food tends to come in large packages—whole or partial animals—while vegetable foods come in small packages, like a nut or a fruit or a leaf. You might spend all day feeding yourself mouthful by mouthful on leaves or fruits, but the leg of a medium-sized antelope might feed several hominids for days. Paleoanthropologists debate earnestly whether early hominids were actively hunting for animals or were scavenging from other predators' kills. Either way, the payoff in terms of high-quality nutrition is enormous if you have the right equipment to be successful. One thing is clear: having stone tools transformed hominid life about 2.5 million years ago.

What difference would a diet suddenly richer in protein and fat make? For one thing, gaining more animal food was probably a prerequisite for the increase in brain size that occurred when early *Homo* evolved from one of the earlier hominids. Brain size is about 80 percent larger in early *Homo* than would be expected for a chimpanzee of the same body size and 20–30 percent bigger than the brains of australopithecines and paranthropines of the same body size.

The issue is not simply that *Homo* had a bigger brain relative to its body size than either *Australopithecus* or *Paranthropus* did—which we, being brainy animals, like to believe is a major advantage. Leslie Aiello of the Wenner-Gren Foundation in New York City and her colleagues have researched the energetic consequences of growing and maintaining a bigger brain. They have put forward an idea that they call the Expensive Tissue hypothesis. They point out that not only is the brain relatively bigger in early *Homo* than in earlier hominids, the body size is also absolutely bigger.

First of all, a bigger body requires more energy, which means either more food or better quality food. Secondly, brains are greedy and consume a lot of energy, both in growth and maintenance. In modern humans, for example, the brain occupies only 2 percent of the body weight but uses about 20 percent of the total energy budget.

How could early *Homo* afford such a big brain? One probable mechanism is a dietary shift fueled by the new opportunities to obtain animal food offered by stone tools. Because meat protein is more digestible than plant protein, an increased proportion of meat in the diet (balanced with increased fat) means that the gut no longer needs to be so long and elaborate. Since guts are full of blood vessels and have more nerve cells than the spinal cord itself, evolving a smaller, shorter gut is a substantial saving of energy. Thus, more meat also means less gut. More meat means having a waist instead of the portly abdominal bulk you see in a gorilla, which needs lots of bacteria in its long guts to break down the masses of cellulose and plant fiber it eats. The advantage of eating more meat is extra energy for growing and maintaining a larger brain, balanced by a lower need for energy to grow and maintain guts. Thus, the increase in relative brain size indicates that nutrition had improved in our ancestors before early *Homo* appeared. Third, the increase in animal protein had to be balanced by more fat, which we know early hominids got by smashing bones open to obtain marrow. The hard evidence and the theoretical energetic requirements jibe nicely.

Hominids transformed rocks into stone tools and stone tools transformed hominids from bipedal apes that are basically herbivorous (plant-eating) into predators. No improvement comes without a cost, though.

Becoming more predatory demanded that hominids pay more attention to other animals than previously. I don't mean "paying attention" in a trivial sense. I mean devoting substantial time and energy to knowing where other animals were, to learning what they were doing and what they were likely to do in the immediate future. From the time that hominids began making stone tools and using them to obtain more meat, fat, hide, and tendon, hominids were subjected to strong selection to improve their observational skills and their memories of what they had observed. They needed to be able to put together observations from various times and places to make judgments about what other animals would do. These mental abilities were probably almost as important to getting animal food as was knowing how to make stone tools.

Of course, other animals also pay attention to each other. It is a truism of animal studies that the behavior of any social species is fundamentally "about" other members of that species. Chimp behavior is about chimps; lion behavior is about lions. Most of the attention an animal expends is focused on other members of its own species, not on other species in the ecosystem. Although lions stare with fascination at antelopes when they are hunting, lions don't actually study particular prey species. The point, to the lion, is that the animal over there is the right size to be prey and moves like prey and smells like prey and is going to be prey if the lion has any say in the matter. Whether it is a warthog or an antelope or a small human being is irrelevant.

To be a better predator than a true carnivore, which has evolved for its role over millennia, humans had to have some extra tricks in their repertoire. And paying attention and making deductions from observations—the beginning of the animal connection—was probably a significant trick from which humans benefitted.

Making a rapid transformation to a predatory lifestyle also posed a serious ecological problem for hominids. It is a fundamental ecological rule that plant-eating herbivores can live at high densities but meat-eating carnivores cannot. Ecologists often classify organisms according to their place in the trophic or energetic food pyramid. Grass and other plants subsist on sunshine, rainwater, and nutrients from the soil. They are abundant and their abundance forms the base of the trophic pyramid. Mammals that eat plants comprise the next step up. These are animals like deer, zebra, elephant, and rabbits. They are much less abundant than their plant resources, but they are still numerous. At the top of the trophic pyramid are meat-eaters—carnivores—which prey on the plant-eaters one step below them.

You can see the trophic pyramid if you think of the seemingly endless grasslands of the Serengeti Plains of East Africa during the great annual migration, when huge herds of zebra, wildebeest, gazelle, and other species move through seeking new pastures. The number of individual grass plants is so large it is very nearly incalculable. The migrating herbivores number in the tens of thousands, if not millions. Yet only a small number of carnivores follow those migrating animals through a particular area: maybe one or two lionesses cooperate in stalking and pouncing on a weakened or sickly antelope.

If the kill is successful, vultures come in to claim part of the kill, perhaps about the time that the entire pride of lions appears to share in the kill. They are followed swiftly by hungry hyenas or packs of Cape hunting dogs, and a few jackals. Every animal wants a piece of the carcass. The lions defend their kill, even the pride's male who usually does not participate in the hunt at all, and the hyenas circle and snap, trying to drive the lions away. The vultures hop around on the ground ludicrously, flapping their wings to look bigger and trying to usurp part of the kill for themselves. Jackals circle nervously at the edges of the skirmish, trying to dash in and grab a leg or a piece of flesh with which they can retreat. In films, a maelstrom of carnivores seem to fight over the carcass, but if you count them, there are probably fewer than thirty individual predators at a carcass compared to the vast herds of antelope and zebra from which they were hunted.

What dictates that herbivores be so plentiful and carnivores so rare? The plants that feed herbivores grow in tremendous abundance, but the transfer of energy from each grass plant to each herbivore involves a net loss of 90 percent of the original energy. Most of the energy in the ecosystem is lost in a single step. The transformation of herbivore tissue into carnivore tissue—when a lion eats a zebra—once again involves a net loss of 90 percent of the energy that was contained in the herbivore. Living at the top of the trophic pyramid involves the risk of going hungry and the risk of injury from struggling prey. Life at the top is precarious and lonely.

Herbivorous hominids who suddenly became predators after the invention of stone tools were in an awkward bind. Remember, this dietary change occurred extrasomatically and very fast relative to an evolutionary change. If hominids continued to live at the same density as when they were herbivores, they would figuratively eat themselves out of house and home. They either needed much more territory—and more animals—for hunting or they needed to have many fewer individuals in their species. Clearly, no conscious decision was involved. How could a herbivorous hominid-turned-carnivore survive?

There are only three evolutionary options that will decrease population density.

One way to diminish the number of individuals on the land is to have a lot of individuals starve. Such a die-off might be difficult to see in the fossil record because every individual who becomes fossilized is dead by definition. However, a paleoanthropologist can see if the number of individuals that suffered from malnutrition or disease (symptoms of a population density that is too high) increased at 2.5 million years ago. During development, the teeth and bones of individuals who are sick or starving develop stress lines that mark temporary cessations in growth. If there were an obvious increase in the frequency of stress lines in the first members of the genus *Homo* as compared to earlier hominids, the starvation hypothesis would be supported. No such evidence has yet been found.

The second possible resolution to this density dilemma is to keep the number of individuals constant but evolve a smaller body size. A small predator needs less food than a large predator and so can make do with a smaller territory. We know this isn't what happened in human evolution either. Though early *Homo* is similar in size to australopithecines or paranthropines, *Homo* soon becomes larger.

The remaining solution to the density dilemma is for the newly predatory species to radically expand its range. Evidence that humans did precisely this shows up unmistakably in the fossil record. *Homo habilis* appeared in East and South Africa about 2.3 million years and *Homo erectus* appeared in the same areas at about 1.9 million years ago. Almost as soon as *erectus* evolved, it expanded out of Africa and into Eurasia. By 1.8 million years ago, *Homo erectus* was distributed from southern Africa to as far north as the Republic of Georgia and as far east as Indonesia and China (Figure 9). As they expanded their range, humans carried with them the knowledge and skills needed to make flaked stone tools and to hunt or scavenge animals, and the ability to observe the other species around them.

Don't think that this territorial expansion happened when some *erectus* consciously looked around and said to his buddy, "Hmm, this place is getting too crowded, let's try someplace new." The more probable scenario is that humans kept looking for more game, which took them into new areas. They were

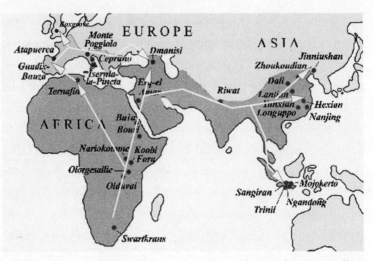

9. *The earliest specimens of* Homo *come from Africa at least 2.3 million years ago. The species then spread rapidly until its geographic range covered most of Eurasia.*

already observing other animals closely because this knowledge gave them an advantage in hunting and in competing with other species. Knowing that game was more abundant to the east or in the valley or nearer to the river would be important knowledge. They probably simply followed the greatest abundance of game. To early humans, as to lions, there was not much difference between hunting an African antelope or an Asian deer. Some humans stayed in Africa, where hominids had always been, and survived with their new skills and abilities. Others drifted to areas where there was less competition and more game, slowly moving northward and then eastward or westward without any deliberate intention of dispersing. In all probability, early humans simply expanded into ecosystems similar to the one they started from but with fewer or no other humans. Humans had no idea that they were leaving the African continent for the Arabian Peninsula or the Arabian Peninsula for Eurasia.

All of these consequences of inventing stone tools sound reasonably positive, but there were also difficulties. Yes, having stone tools made humans into effective predators, improved their diets, enabled them to enlarge their brains, and forced them to expand their geographic range. Having stone tools and changing their ecological niche like this also put humans into deadly danger.

As scavengers and predators, humans became direct competitors with the true carnivores. A carcass may yield lots of good food but it attracts some ferocious competitors who will fight mercilessly for access to that food. Power scavenging is just one aspect of a set of aggressive behaviors between species known as "interference competition." It is very dangerous.

A lot of predators die because of interference competition over carcasses and a lot more avoid areas where interference competition is too common. After wolves were reintroduced into Yellowstone National Park, for example, a study of coyotes and wolves found that wolves were the dominant cause of death for transient coyotes (as opposed to long-term resident coyotes). As a result of the reintroduction of wolves, coyote density declined by 39 percent in the park. Once the big dogs moved into the area, the little dogs had a much harder time.

How serious would the risks of interference competition have been for early humans in Africa? Very.

Today, there are six medium- or large-sized carnivores (bigger than about 45 lbs or 20 kg) in African habitats: lions, spotted hyenas, brown hyenas, striped hyenas, leopards, cheetahs, and

Cape hunting dogs. Lions and spotted hyenas are the largest and generally predominate over the smaller species in confrontations, which are often decided by body size. These largest African carnivores are also social, which gives them an additional advantage. Today, unarmed humans can drive lions or hyenas from their kills, but it is a risky strategy. The humans are more likely to lose the carcass, or their lives, than the lions or hyenas are.

The situation was grimmer in the past. Blaire Van Valkenburgh of the University of California, Los Angeles, has been studying carnivore guilds—groups of animals that live in the same habitat and have similar ecological niches—for her entire career. She says that when hominids first started making tools about 2.6 million years ago, there were eleven large carnivores, not six: lions, leopards, cheetahs, striped hyenas, brown hyenas, an ancestral spotted hyena, a running, long-legged hyena, a wolflike predator, two species of sabertooth cat, and a false sabertooth cat (Figure 10). False sabertooths are a different evolutionary lineage from the true sabertooths but share the characteristic of having very long, formidable canine teeth.

In fact, a false saber-toothed cat was also around in *Ardipithecus*'s time and is my favorite candidate for the predator that ate the female individual nicknamed Ardi.

The high number of carnivores at the time when hominids were first making tools meant that interspecific competition among predatory species was particularly intense. Eight of the eleven carnivores outweighed humans. Some of the extinct species, like the sabertooth cat—*Homotherium*—weighed five times as much as early *Homo*. An individual human would probably have come out badly in a face-to-face competition with any of these ancient carnivores. Just possibly working in groups,

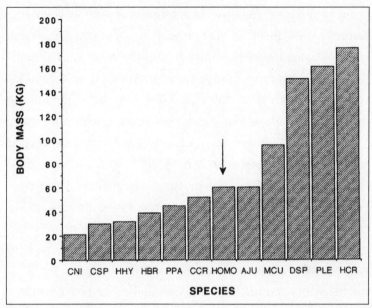

10. *About 2.5 million years ago, eleven carnivores in East Africa competed for prey with humans (arrow). Humans weighed no more than a medium-sized carnivore and were probably driven away from carcasses by larger carnivores. The abbreviations are:* CNI = Chasmoporthetes nitidula, *a running hyena;* CSP = Canis indeterminate wolflike species; HHY = Hyaena hyaena, *the striped hyena;* HBR = Hyaena brunnea, *the brown hyena;* PPA = Panthera pardus, *the leopard;* CCR = Crocuta crocuta, *the spotted hyena;* HOMO = early *Homo;* AJU = Acincononyx jubatus, *the cheetah;* MCU = Megantereon cultidens, *a saber-toothed cat;* DSP = Dinofelis species, *a false saber-toothed cat;* PLE = Panthera leo, *the lion;* HCR = Homotherium crenetidens, *a saber-toothed cat.*

throwing stones, and having stone tools might have tipped the balance in favor of humans on some occasions. But early humans would have had to watch their backs, especially if they were in the vicinity of an attractive carcass.

Between 2.3 and 1.7 million years ago, humans increased in body size from about 70 lbs to about 120 lbs (35 kg–60 kg), judging from their bones. This increase in body size may have been an adaptive response to intense competition from true carnivores. By about 1.7 million years ago, early African *Homo erectus* was no longer the "little guy" in the ecosystem but outweighed at least four of the African carnivores. Two other carnivores had already gone extinct, possibly because of the intense interference competition. Bigger body size was a big advantage.

Even with a bigger body size, predatory humans needed social groups, tools, sharper sensory skills, and a more intense focus on other animals than they had had as herbivores. Humans needed to pay careful attention to the animals around them. These traits were useful in finding and killing prey and in watching warily for the other carnivores that were likely to arrive as soon as the prey was downed. An enhanced focus on the behaviors and habits of other animals would have an obvious payoff for hominids in terms of improving their diet and keeping them alive. Certainly, the enlarged brain of early *Homo* was important in paying close attention to other species and in storing information about those species for future use.

The evidence of the fossil and archeological record is clear: our deep involvement with animals began over 2.6 million years ago and was facilitated or even necessitated by the invention of stone tools. The behavior that underlies the beginning of the animal connection is paying attention to other animals, learning about them, and their habits and needs. Stone tools changed not only our ancestors but their entire ecosystem. The enormous evolutionary advantage of being a brainier hominid that had learned to gather information about other species is obvious.

Was 2.6 million years ago really the magic moment when the first hominid figured out how to make tools? Maybe, or maybe not. If any hominids were making tools earlier than that, there are three possible explanations why no one has found them yet. First, those tools could be so rare that none has yet been found. Second, those tools could be so crude or so little modified as to be unrecognizable. Or, third, the pre–2.6-million-year-old tools could have been made out of perishable materials (not stone) that do not usually survive.

4

Is a Tool a Tool a Tool?

UP TO NOW, the story sounds good and my hypothesis is well supported by a lot of evidence. But there's another complicating factor muddying our understanding of early technology. We know stone was not the only raw material that could have been used to make tools.

Couldn't there have been tools made of twigs, sticks, leaves, other natural objects? Yes, there could have been, but we don't have any of them until almost half a million years ago, when there is a wooden spear from Clacton-on-Sea in the United Kingdom and three wooden spears at a West German site near Schöningen. Some paleoanthropologists argue that early hominids probably did make tools out of plants, leaves, branches, and twigs, since some apes, monkeys, and other mammals and birds make such tools today. Analogy to other species makes such an idea seem plausible, but my "plausible" may be your "incredible."

Starting roughly 2 million years ago—or 600,000 years after the oldest stone tools were made—what we do have are bone tools. There are strong hints that these special tools were made and used by paranthropines, whether or not they used stone tools. Between 2 and 1 million years ago, bones or broken pieces of bone were selected for use but they were not modified much, if at all, to make them suitable for use. We can show that the hominids preferred bones with particular attributes, such as particular skeletal elements (limb bones), size, density of the tough outer bone, shape, and degree of weathering. The minimal modification means that early bone tools—bone expediency tools—are tricky to identify.

The difficulty is not for want of suggestions. Since the nineteenth century various paleoanthropologists and paleontologists have claimed that broken bones associated with hominid remains were tools. Raymond Dart, a pioneering and imaginative South African anatomist, even coined the term Osteodontokeratic (bone-tooth-horn) culture for peculiarly broken specimens he excavated along with some of the first australopithecine remains. Unfortunately, his claims, like those of many predecessors, were weakly supported by objective evidence. It also worked against him that his writing style sounds more like the words of an evangelical preacher than those of an objective scientist. It is all too easy to look at a broken, fossilized bone and think up something the specimen might have been useful for. Proving a bone has been used is a different matter. Skepticism about early bone tools was rampant from about 1950 onwards.

Paleoanthropologists were surprised that a hardheaded scientist like Mary Leakey designated 125 specific bones from the Olduvai Gorge in Tanzania as tools in her definitive book on Olduvai in 1971. She based her assessments on her extensive experience with tens of thousands of fossils and stone tools from Olduvai that were made between 2 and 1 million years ago. What she had was a keen eye for an anomaly—for something different—but she wasn't interested in spending much time studying natural and artificial causes of bone breakage and modification. Frankly, many skeptics, some of whom had never seen the original specimens, were unconvinced by her assertions and others simply let the matter drop.

When Rick Potts and I first noticed cutmarks on one of the bones from Olduvai, we used a microscope as a means of documenting and identifying bone modifications. Our instrument of choice was the scanning electron microscope (SEM) because it made high magnification possible and, even more important, allowed us not only to see bone surfaces in three dimensions but to record them. (Light microscopes tend to visually flatten objects put under them and make it harder to assess three-dimensional topography.) In the 1980s I decided to follow up on our studies of cutmarks and other surface modifications on bones by applying a similar technique to Mary's putative bone tools. By then I had comparative collections of bones modified by many natural agents and processes—various animals, weathering, abrasion, and plants—that showed what natural, nonhuman processes did to bones. I added new experimental bone tools that I or colleagues made and also examined many bone tools made and used by nonindustrial peoples, like the historic Plains Indians of North America.

After detailed examination of Mary's putative bone tools and the comparative samples, I concluded that many of the bone tools she had identified bore microscopic traces that looked like the use wear on the comparative and experimental samples. A key point was that the wear and polish were confined to within a small distance of the working edge of the tool; the whole specimen was not worn or polished, as happens when a bone is abraded in a river, damaged by weathering, or altered by trampling, for example.

All of the Olduvai specimens I reidentified as bone tools looked very fresh—unweathered—and they were mostly from very large animals: hippos, elephants, giraffes, and the like. Some of the Olduvai specimens showed overall abrasion and I had to classify those specimens as ambiguous, because the abrasion obscured whatever might have lain underneath. Most of the Olduvai bone tools showed signs of the deliberate removal of bone flakes, using knapping or flaking techniques very similar to those used to make the earliest stone tools. My statistical analysis of the bone tools identified from their microscopic use wear indicated that, compared to unused animal bones from the same sites, significantly more flakes had been removed from the bone tools than for other Olduvai bones matched for body part and species.

Hominids had been flaking stones since 2.6 million years ago, so realizing that the same techniques could be applied to the bones of very large animals that had thick cortical bone—such as leg bones—wasn't a great imaginative leap. The differential use wear on the bone flakes provides a direct confirmation of use; knowing a bone has been flaked does not mean necessarily that it has been used. I think probably most of the flaked bones were used as cutting tools, as stone flakes were, even though I can't always prove it.

Having tried to use bone flakes for cutting meat, I am unimpressed with them as cutting tools. They cut reasonably well initially but do not hold a sharp edge for very long.

The edge soon gets clogged with bits of connective tissue or meat that makes cutting more difficult. Frankly, I don't know why someone would use bone flakes if stone flakes were available—and all the sites with bone flakes also have stone flakes. Perhaps enough stone flakes were not available; perhaps the hominids were in a hurry to cut off some meat and get away from the carcass, so they used whatever they could get their hands on. I don't know. But I know that somebody did use those flakes for some reason, and until I can think of another way to gain more information, that conclusion will have to suffice.

The most remarkable bone tool at Olduvai is a piece of elephant bone flaked on both sides to a teardrop shape just like the classic stone tool from Olduvai often called a handax, which is the signature of the Acheulian industry (Figure 11). I

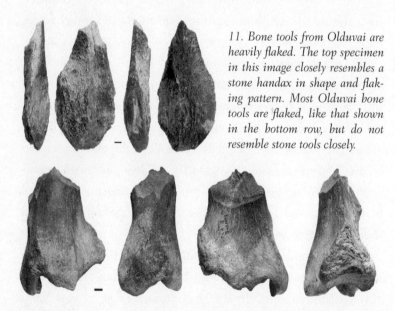

11. Bone tools from Olduvai are heavily flaked. The top specimen in this image closely resembles a stone handax in shape and flaking pattern. Most Olduvai bone tools are flaked, like that shown in the bottom row, but do not resemble stone tools closely.

found it amusing to show this specimen to some highly skeptical archeologists. If they see the specimen from across a room, for example, they immediately proclaim it to be a handax. One archeologist, once he got closer and saw that the raw material was bone, clearly wanted to recant his identification but couldn't deny the telltale shape of the specimen. "There's only *one* of them, though," he backpedaled. "How do we know the maker knew what he was doing?'

Indeed, there is only one bone handax at Olduvai and we have no idea what its maker thought. But that telltale shape and pattern of flake removal was no accident, to my way of thinking. (There are several other beautifully fashioned bone handaxes from other, much later sites, like Castel di Guido in Italy, that must have been made by a different hominid species. They are remarkably similar to each other.) Again, I can't imagine that a bone handax would have been effective as a tool for long, but the specimen is unmistakably a handax—a bifacially flaked tool made by the same process that produces a stone handax—with use wear on its pointed tip.

Four fascinating specimens from Olduvai show a series of indentations made by the same punch or hammer, which was probably a stone tool, on a natural flattish surface (Figure 12). Originally, I thought these were anvils, but later work by friends and colleagues Lucinda Backwell of Witwatersrand University and Francesco d'Errico of the University of Bordeaux on the same specimens suggested these bones were more likely used as hammerstones. Lucinda and Francesco experimented with using such bones as hammerstones and were able to reproduce very similar marks. But photos of crocodile damage published later by Jackson Njau and Rob Blumenschine suggest to me

12. *(top) These bisected punctures on an elephant kneecap may have been made by hominids' using the bone as an anvil or hammerstone. Viewed in close-up, these marks also resemble (left) bisected punctures made by crocodiles, so more work is needed.*

that these marks might have been made by crocodiles, so these "hammerstones" or "anvils" will warrant another look.

At this point, my studies had gone a long way toward convincing paleoanthropologists that there were simple bone tools early in the archeological record and that detailed examination of the working edges of those tools could confirm their use. The data showed that East African hominids were preferentially knapping the bones that had thick, dense tissue on the outside, known as cortical bone. Mostly they selected the leg bones of hippo, giraffe, and elephant. It seems that the hominids transferred the familiar knapping technique from stoneworking to boneworking and were also using intact bones of suitable shapes as hammerstones. Both types of bone tools were much much less common than stone tools or unused bone tools.

After I presented my studies of the Olduvai bone tools at a conference, I was contacted by a brilliant South African researcher named C. K. Brain. Known to everyone as Bob, he is one of the cleverest and kindest researchers it has ever been my pleasure to work with. He has a scientist's nose for evidence and a natural historian's understanding of the real world. He brought to my lab—then at Johns Hopkins Medical School in Baltimore—a set of possible bone tools from two famous hominid cave deposits in South Africa. The cave yielding the majority of possible tools—Swartkrans in Gauteng Province—had also yielded mostly *Paranthropus* remains, with some specimens of *Homo* in particular layers. The other cave, Sterkfontein, also in Gauteng Province, had a few possible bone tools and mostly *Australopithecus africanus* specimens. Along with stone tools and hominid remains, these caves both yielded thousands of animal bones. Bob's bones were between 1.8 and 1.1 million years old, making them contemporaneous with the Olduvai bone tools I had studied.

To my surprise, Bob's specimens looked nothing at all like the bone tools from Olduvai (Figure 13). His specimens were not flaked, nor did any show depressions suggesting use as an anvil or hammerstone. What he had was a series of splinters of animal long bones, each of which had a tip that was rounded and looked worn, instead of being sharp, like a freshly broken bone. I have never seen anything like this in the vast Olduvai collections.

Together, we made high-fidelity replicas of the working ends of the specimens and looked at them under the SEM, as I had the Olduvai bones. We found clear signs of use wear on many of these specimens. Tellingly, the wear occurred on the tips and continued for only a few millimeters down the shafts, but the apparent polish did not cover the specimens. On top of the pol-

13. *Bone tools from Swartkrans, South Africa, are roughly contemporary to those from Olduvai, but are not flaked and look very different.*

ish we could see a mixture of fine and coarse scratches. After we had studied and photographed the replicas, we carried out parallel studies on some bone tools that Brain and his son Nad had used experimentally to dig for tubers in the hard South African soil. We also compared the fossils to additional specimens from my comparative collections.

We could tell the Swartkrans specimens had been used because they showed clear differential use wear. The closest matches in our reference collections were the bone tools used to dig for tubers and roots in the rocky soil surrounding the caves. We concluded that these pieces of broken bone were the equivalent of digging sticks, a common tool still used to dig for roots or tubers by living peoples in arid areas. The finding made sense given the paucity of trees and useful branches in the dry savannahs of South Africa.

Roughly ten years after Bob and I finished our work, our good friend Lucinda Backwell of the University of Witwatersrand took another look at the Swartkrans and Sterkfontein bone tools. From an extensive study of all 23,000 of the Swartkrans fossils, Lucinda identified another 16 possible bone tools. She also compiled an extensive reference collection of 13,000 bones, much larger than the one Bob and I had used to evaluate the use wear on the South African specimens. Among her experimental bone tools were ones she had used to dig for tubers, others she had used to pierce and open termite mounds to get at the termites within, and tools she had used to pierce and scrape animal hides. Collaborating with Francesco d'Errico, who had earned a strong reputation for his microscopic analyses of fossil bones, Lucinda compared the combined Swartkrans and Sterkfontein materials to the new and larger reference collection. They used the SEM to document and study the use wear, and they had the advantage of image analysis programs that hadn't existed when Bob and I did our studies.

Their results were exciting. Like us, Lucinda and Francesco found a pattern of highly localized polish overlain with scratches. Where the pair surpassed our work was in being able to measure the size and orientation of the numerous scratches on the working ends of these tools and on bones from the reference collections, so that the comparisons could be made statistically and not simply by visual inspection. (Visual inspection is a jargon term for "eyeballing" the specimens.) Their metric and statistical procedures proved that the ancient bone tools most closely resembled termiting tools, not bones used for digging up tubers as Bob and I had thought (Figure 14).

They also considered the question of whether hominids chose any handy bone for use as a tool or had specific crite-

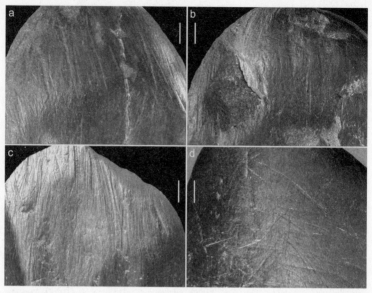

14. *Microwear on the fossil bone tools from (a) Swartkrans and (b) Drimolen in South Africa more closely resemble microwear on (c) experimental tools used to break open termite mounds and extract termites than (d) experimental bone tools used to dig up tubers.*

ria in mind. They measured the bone tools and compared them to nontool bone fragments from Swartkrans, again finding an important difference. The bones chosen as tools were longer, wider, and had thicker shafts than the unused bone fragments. Nonetheless, these bones were not from very large or extremely large mammals, like the Olduvai bone tools. The Swartkrans and Sterkfontein bones tended to come from medium-sized antelopes.

Before this marvelous study in 2001, no one had envisioned early hominids as eating substantial quantities of termites. And yet, we know that our close relatives the chimpanzees eat termites and ants, and we know that these insects were present at

Swartkrans because the site also yielded the remains of special-
ist myrmecophages (ant- and termite-eaters) such as aardvarks,
aardwolves, and pangolins. Bob and I had simply never even
considered that the bone tools might have been used to dig for
termites and we had no samples of bones that had been used in
that way. We were both thrilled by the results of Lucinda and
Francesco's study.

Their work in turn raised the same old question: Who made
the bone termiting tools?

Both early *Homo* and *Paranthropus robustus* lived in South
Africa between 2 and 1 million years ago. However, only the
bones of *Paranthropus* have a telltale chemical signature (a high
proportion of a particular isotope of carbon) showing that this
species had eaten a surprising amount of protein during its life.
Eating a lot of meat would give bones this isotopic signature, of
course, but so would eating lots of termites.

Termites are surprisingly nutritious—roughly twice as nutritious as ants. Even more surprisingly, Lucinda told me, "One
hundred grams of rump steak yields 322 calories, but 100 grams
of termites yields an amazing 560 calories." (One hundred
grams, or 2.2 ounces, of termites equals about 250 individual
termites. An aardvark that weighs about as much as a paranthropine may eat as many as 40,000 termites a night.) Not surprisingly, the bones of aardvarks from Swartkrans had very similar
carbon isotope values to those of *Paranthropus*. So, although
Paranthropus had the teeth of a vegetarian, this analysis revealed
that it had the bone chemistry of one that ate lots of protein.
Similarly, chimps have teeth suited for eating vegetable foods
(mostly fruit), but may also eat lots of termites. Perhaps termites
were an important fallback food, eaten intensively by paran-

thropines in the dry season when vegetable foods were scarce. There are no anatomical signs that paranthropines were specialists committed to eating ants and termites, which would have made them myrmecophages like the aardvarks, aardwolves, and pangolins. But plenty of animals that are not myrmecophages eat termites as a small yet important part of their diet.

Lucinda and Francesco expanded their research program to reexamine the Olduvai bone tools. They found nothing in the Olduvai collections resembling the South African bone termiting tools. They confirmed that there were bone tools at Olduvai, as Mary Leakey had always claimed, and that they were heavily flaked compared to other bones, and usually derived from very large animals with thick cortical bone.

Then, in 2008, Lucinda and Francesco investigated twenty-two possible bone tools from a third South African hominid site in Gauteng Province, Drimolen, dated to between 2 million and 1.5 million years ago. Eight of the specimens did not have the microscopic characteristics of bone tools and were simply rounded bone splinters. However, fourteen had wear patterns that matched the experimental termiting bones and the fossil termiting bones from Swartkrans. The types of bones chosen for use at Drimolen and the length of the splinters matched those from Swartkrans as well.

Apparently, fishing for termites using fragments of long bone shafts was a widespread and enduring pattern in South Africa, but not in East Africa, between about 2 and 1 million years ago. But exactly who was using them? Again, both *Homo* and *Paranthropus robustus* were present at Drimolen, yet *Homo* is represented by only two specimens, and *Paranthropus* is overwhelmingly dominant, with seventy-seven specimens. Since the bone tools at Swartkrans are also more common in the geological layers that contain *Paranthropus* but lack *Homo,* the clear implication is that *Paranthropus* was more likely to have been making and using the bone tools.

Because *Paranthropus* was sexually dimorphic, the males much larger in body size than the females, they are sometimes thought of as being like gorillas. Modern gorillas don't use tools to go termite fishing, but chimps do, so the idea that a gorillalike hominid might have used crude tools to collect termites doesn't seem so far-fetched. Lucinda and Francesco even speculate that if *Paranthropus* were using bone tools for termiting, the behavior may have been performed only by females and juveniles, as is the case among chimpanzees.

I am fascinated that the bone tool traditions of South and East Africa are so very different while the stone tool traditions in the two areas are fairly similar. East African bone tools included a few hammerstones but were mostly flaked bone, akin to the flaked stone tools of the region. It is as if the East African hominids learned a technique and then tried it out on various substances to see what worked. In South Africa, the bone tools are not flaked, do not come from very large animals, and are splinters from much smaller animals that were used as probes or digging sticks in a manner totally different from the use of flaked stone.

Why didn't South African hominids flake bone?

Why didn't East African hominids use splintered bone probes? Were they using twigs or vines instead of bone fragments for termiting? Or weren't East African early hominids termiting at all?

And was the same hominid making bone tools in both places? The evidence that the South African bone tools were

made by paranthropines is intriguing but not ironclad. Paranthropines were also present at Olduvai when the bone tools
were being flaked, although traditionally most paleoanthropologists assumed *Homo* was making the bone tools. But maybe
bone tools are a paranthropine way of making tools in imitation
of early humans. Or possibly paranthropus made both the bone
tools and some of the stone tools too.

Is this difference between the eastern and southern technologies a sign that the East and South African hominids had
little or nothing to do with each other—that they were inventing
tool technologies in isolation from each other?

Do the differences in bone tools have anything to do with
the fact that *Homo* survived and paranthropines and australopithecines went extinct?

These questions continue to haunt me. I have no answers
yet, only questions.

5

Uniquely Human?

IT'S TIME TO STEP BACK from the details of the invention of stone tools and bone tools and their ecological impact to take a broad view of toolmaking.

The traditional view has been that toolmaking and tool using are behaviors unique to the human species. This view is incorrect, although tools and extrasomatic adaptations have been of great importance in human evolution. Charles Darwin proclaimed that only humans make and use tools, following the ideas of even earlier thinkers, but he was wrong in this case. A classic book in anthropology is *Man the Toolmaker*, written in 1949 by Kenneth Oakley of the British Museum of Natural History. Oakley's point was that the ability to make tools at all was what made that man a human.

In the rather formal language of the day, Oakley stated that "Man is a social animal, distinguished by culture: by the ability to make tools and communicate ideas. Employment of tools appears to be his chief biological characteristic, for considered functionally they are detachable extensions of the forelimb. . . .

Relying on extra-bodily equipment of his own making, which could be quickly discarded or changed as circumstances dictated, man became the most adaptable of all creatures."

But Oakley, too, was wrong. Toolmaking is not an exclusively human trait. Many animals have been observed making tools of various kinds in the wild since at least the nineteenth century. Mud wasps hold pebbles in their jaws and use them to tamp down the mud they make their nests from. Sea otters use rocks to break open shellfish. Egyptian vultures throw rocks at eggs to break them open, while finches and crows use twigs or cactus spines to pry insect larvae out of crevices. Elephants use branches as fly whisks and chew bark to use as sponges. Capuchin monkeys use rocks to break open nuts.

The arguments about toolmaking and tool using came to the fore in 1964, when Jane Goodall published a groundbreaking paper in *Nature* in which she documented her observations of chimpanzee toolmaking. Although some reports of animal tool using preceded Goodall's observations, her work was perceived as a major challenge to the separation between humans and the rest of the animal kingdom. Louis Leakey, Goodall's mentor, is said to have chortled when he got the news of her discovery, "Now we must redefine tool, redefine Man, or accept chimpanzees as humans." The position that chimpanzees are humans has not garnered much support.

So, why hasn't toolmaking transformed chimpanzees the way toolmaking transformed some hominids?

Chimpanzees are among the most adept nonhuman toolmakers in the wild. They use leaves as sponges, modify twigs to go termite fishing or to obtain honey, pound and smash nuts with branches and rocks, sharpen sticks with their teeth to use as spears for hunting bushbabies in their tree holes, dig for tubers and roots using sticks and, similarly, use sticks to extract marrow from broken long bones too narrow in diameter for their fingers. Fifty-six different behaviors that might be considered tool use in the broadest sense have been documented as varying in frequency among the seven best-studied chimpanzee populations in the world, though many of these behaviors are functionally very close to others (such as ant fishing versus termite fishing).

Is chimp tool use really comparable to Oldowan tool use?

Recently, a group of researchers led by Julio Mercader of the University of Calgary excavated what they called the world's first known chimpanzee archeological site. It is comprised of three small stone assemblages (206 stone tools or pieces of broken stones), most of which came from a single site, Noulu. The site is in the Taï National Park in the Côte d'Ivoire and is dated to 4,300 years ago. Detailed, long-term studies of chimpanzee behavior have been carried out in the Taï forest, so we know that the local chimps used hammerstones and rock anvils to smash open panda nuts, among other things. The key question of the research project led by Mercader was: Was the archeological site they found in the Taï forest formed by chimps or by humans?

First of all, Mercader and his colleagues wanted to be sure archeologists could tell stone tools from natural mimics of stone tools. They presented three archeologists with a blind test involving ninety stones. One third of the objects had been naturally fractured by geological forces in Canada; one third was from the Taï forest site and had been selected as most probably showing signs of deliberate flaking; and one third came from a

15. *These stone fragments from Noulu are interpreted as being broken by chimpanzees during nut-cracking activities.*

human archeological site in Canada. Could the archeologists tell the difference?

They had no difficulty distinguishing the naturally broken stones from the real stone tools and agreed closely on their choices. They specified twenty-eight specimens that had been deliberately flaked, whereas another thirty-five had been broken unintentionally during bashing activities (Figure 15). Only the flaked stones from the Canadian archeological site and a few of the Noulu specimens showed conchoidal fractures. Lacking conchoidal fractures, nearly all of the specimens from Noulu were identified as unintentional byproducts of "thrusting percussion" (nut-bashing). Microscopic starch grains from those nuts were actually found on some of those specimens.

So, who made the tools? Did chimps or human inhabitants of the forest produce the broken and battered specimens from Noulu?

About the flaked stones with conchoidal fractures, there is little doubt. Chimpanzees have never been seen to flake stone intentionally. As Mercader and his colleagues remark, perhaps wistfully, "the possibility that humans could be the sole culprit of our stone collections must be carefully examined."

Clues as to the identity of the maker of the other, accidentally broken stones come from estimates of the sizes of those stones prior to breakage. Because chimpanzees are much stronger than humans and have larger hands, they prefer to use larger stones as tools. For example, hammerstones from Oldowan assemblages made by early hominids usually weigh less than about 1 lb (400 g), with a maximum of about 3 lbs (1,000 g). Hammerstones used by Taï chimpanzees today are much heavier. In a collection of 133 hammerstones from modern chimpanzee nut-bashing sites, 65 percent weighed 3–20 lbs (1,000–9,000 g) and the heaviest was a staggering 53 lbs (24,000 g). Using these figures for reference, Mercader's team deduced that the Noulu stones, which average 1.5 lbs (710 g) in weight, are more likely to have been made by chimpanzees than humans.

The team therefore concluded that they had found the first chimpanzee archeological site, even if some of the specimens were flaked by humans. This means that chimpanzees and hominids share certain cultural attributes. They transport stone from one place to another for future use; they gauge the suitability of the size, weight, and material properties of the stones against the intended use; they return and reuse particular spots (favored locales?), which allows stones and other debris to accumulate in a concentration; and they create activity areas where a particular task is carried out.

These shared behaviors are extremely interesting, but they do not indicate that chimpanzees were making tools as early hominids did.

Hélène Roche of the Centre National de la Recherche Scientifique in Talence outside Bordeaux is one of the foremost experts on the Oldowan, having found, excavated, and analyzed 2.3 million-year-old sites at Lokalalei in northern Kenya. Hélène and her CNRS colleague Anne Delagnes are unimpressed with the Noulu chimpanzee tools. They observe critically that hominids making early Oldowan tools had already far surpassed "the stage of an unintentional production of debris such as that resulting from the accidental breakage of hammerstones at the nut-cracking loci of chimpanzees . . . which should not be mistaken for intentional flaking."

Other experts in Oldowan technology, like Nick Toth and Kathy Schick of the Stone Age Institute, state their opinion even more bluntly. The stone debris from Noulu "does *not* mimic Oldowan occurrences." (Italics in original.) To their eyes, the flaked stones were humanly modified whereas the broken ones weren't, and the difference was clear.

Physically, a chimp placing a hard nut on a sturdy stone anvil and bashing it with a hammerstone performs an action similar to what an early hominid did to flake a core. The hammerstone or the anvil may shatter, creating sharp fragments that could be very useful. However, the hominid deliberately striking a core is performing an activity that is not aimed at extracting a nut but is designed to create a conchoidally fractured flake. Flaking requires much greater manual dexterity than nut-bashing; and to be performed correctly, it requires an underlying, complex knowledge of the way rock fractures. Flaking is much more than using an external object (a hammerstone) for a purpose (breaking a nut).

Flaking means using a tool to make a tool, a behavior which attests to a second level of technology more advanced than that of simply using fortuitously shaped objects.

This activity is sometimes termed *metatool making* or *secondary toolmaking*. To the archeologist Sophie de Beaune of the University of Jean Moulin–Lyon III and others, the difference between using an object as a tool and using an object to make a tool is enormous:

> The moment when a hominid or one of its immediate ancestors produced a cutting tool by using . . . percussion . . . marks a break between our predecessors and the specifically human. . . . Which is the contribution that most clearly identifies hominization—the conception of tools not found in nature (which is within the reach of chimpanzees) or the steps allowing the resolution of the problem of cutting by the production of cutting edges?

By "hominization," de Beaune means the process of evolving the distinctive traits of the hominids: becoming more human.

She clearly sees the production of new tools—tools that transform the physical properties of the blank or source material—as essentially different from using slightly altered raw materials as tools.

This is one of the fundamental differences between hominid toolmaking and chimpanzee toolmaking. From the outset, hominid toolmaking reveals an understanding that the obvious attributes of an object can be transformed. In contrast, chimpanzees in the wild select as tools objects that already have the appropriate attributes, whether those attributes are being long and slender (for termiting, say, or spearing bushbabies) or dense and sturdy (like the stones used for bashing nuts) or flexible and absorbent (like the leaves used to soak up liquids). Chimps do not make tools by transforming the properties of the raw object but only by further accentuating them, for example, removing leaves from termiting twigs or sharpening the end of a spear.

Like hominids, chimpanzees and capuchin monkeys often use tools to obtain food, most frequently as anvils or hammerstones to break open hard nuts. However, nonhuman primates do not use tools to obtain animal foods except in three specific cases: termiting; bushbaby spearing; and marrow extraction.

Termiting, though technically a means of obtaining animal food, is not radically different from what hominids probably did with bone termiting tools. Colonial termites or ants live in large nests—mud apartment complexes, more or less. Chimps make holes in the mud nests, then take vines or slender twigs, strip them of leaves, and insert them into the openings. The insects swarm, grab onto the tool, and can be gently pulled out by a chimp, who then delicately nibbles the insects from the stick, as if the whole thing were a kebab. Obtaining insects is quite a different task from hunting mammals. The body size of each termite or ant is very small, as if it were a berry or nut, and the risk of the insect's striking back or evading capture is minimal. Termiting differs from hunting in the normal sense of the word, and termiting sticks are not really hunting tools.

However, spears are certainly hunting tools. Jill Pruetz of Iowa State University and her team observed chimpanzees in Fongoli, Senegal, sharpening long sticks and thrusting them into tree holes, sometimes striking and injuring or killing a sleeping bushbaby. When chimps withdraw the stick, they lick and sniff the end, to see if the tool has pierced a bushbaby. If it has, they reach in, extract the bushbaby, and eat it. There are telling differences, though. Fongoli chimps use the spears for hunting, but they do not cut up bushbabies or other prey with tools; they simply eat prey directly, using their hands and teeth. Chimps can and do catch bushbabies by hand as well, so the spear does not provide access to a new food. What the spear does do is save them from putting their hands into tree holes inhabited by who-knows-what. Although spearing is clearly "about" hunting, in another sense it is more similar to sticking twigs into termite nests to extract termites or into beehives to extract honey than to hunting per se. Chimpanzee spears facilitate the extraction of hidden and high-energy resources.

Chimps also use twigs to pry the marrow out of the bones of small prey animals such as colobus monkeys or newborn gazelles. The long bones are broken by hand or by chewing on them, but the marrow cavity is often too narrow for a chimp's finger to fit inside. A twig or stick is used to extract the marrow in these cases. This use, too, is basically a fishing or extraction operation, and the twig is not a killing or butchery tool.

After decades of observations at seven different long-term study sites, we can pinpoint some key ways in which chimp tools differ from hominid ones:

- Very few chimpanzee tools are used for hunting, though many are used for obtaining food.
- No chimp tools are used in butchery.
- No chimp tools involve flaking stone, though some may involve breaking stone.
- Chimps don't use tools to make other tools.

Though chimps seem to possess the requisite cognitive and physical abilities for toolmaking, their performance as makers of stone tools and metatools falls well short of early hominid behavior.

Was this difference in performance the reason that the invention of toolmaking changed hominids so much and chimps apparently so little? Or did the crucial difference lie in the purpose for which tools were made?

6

The Bonobo Solution

WHO BETTER TO ASK about toolmakers' intentions than one of the world's foremost experts on nonhuman toolmaking, Kanzi the bonobo?

Kanzi of the Great Ape Trust in Des Moines, Iowa, has been working with Nick Toth and Kathy Schick of the Stone Age Institute for twenty years, making flaked stone tools. What is particularly interesting about Kanzi as a toolmaker is that bonobos rarely make or use tools in the wild, compared with chimps who make and use many different types of tools. But bonobos are intensely social—much more so than chimps—and this difference in temperament might account for the fact that bonobos have been much more successful than chimpanzees at learning all sorts of tasks in captivity.

Nonetheless, Sue Savage-Rumbaugh, a scientist with special standing at the Great Ape Trust who has worked with Kanzi for his entire life, initially had doubts about the project's potential for success. At the beginning of the collaboration with Nick and Kathy, Sue had been studying Kanzi and working with him for ten years. She didn't say so, but she didn't believe the bonobos would be interested in knapping and thought that toolmaking might just be too advanced for them, especially since she and the other caretakers found knapping a difficult skill to learn. It is to her great credit as a scientist that Sue did not dismiss the idea of teaching Kanzi to knap as soon as Nick and Kathy suggested it, given her intuitions. Some of her foreboding was certainly borne out.

Both Nick and Kathy are expert knappers, but, frankly, Kanzi was not a natural at the task. He was not even interested in it. As Sue anticipated, the first problem was to motivate Kanzi to try to flake stone—to get him to pay attention to the process. Paying attention is a more important skill than most people realize. To help motivate Kanzi, Nick and Kathy constructed a box that could be opened only by cutting a cord; then they put grapes inside the box while Kanzi was watching. Food is a wonderful motivator for bonobos. Then Nick sat outside Kanzi's enclosure—in clear sight—made a stone flake, cut the cord, opened the box, and generously handed the grapes to Kanzi.

After Kanzi had watched this procedure a few times, the box with grapes was put into his enclosure, along with rocks. As Sue tells the story, Kanzi showed no inclination to pick up the rocks and flake them, so she put rocks in each of his hands and encouraged him to try. Rather reluctantly, Kanzi brought the two rocks together horizontally in front of his chest with very little force. Of course Kanzi did not strike off a flake because he was not hitting precisely and hadn't observed anything about the striking platform or even that there had to be a striking platform for knapping to work.

Kanzi's rocks did not fall apart in the magical way that Nick's had, and Kanzi quickly concluded that he couldn't make a flake. Nick and Kathy continued to demonstrate, trying to tempt Kanzi into cooperating by putting extra special treats in the box. After his initial failure, Kanzi simply refused to try. Sue thought Kanzi was sensitive to failure and resented being asked to do something that was impossible for him. Eventually, he hit the rocks together more forcefully and made a tiny stone flake. This small success kept him going in his attempts.

Though Kanzi understood what he was being asked to do, execution of the task was especially difficult because of bonobo anatomy. His hands are too large, his fingers are too long, and his thumbs are too short to grip the core and hammerstone and effectively strike a flake as Nick does. Kanzi discovered his physical limitations very quickly.

In time, Kanzi found his own solution to the problem. After about two weeks, he picked up a stone in his right hand, rose to a bipedal stance, and hurled the stone with great force against the tiled floor. The stone shattered into many sharp and useful pieces—more than Kanzi had produced up to that point in all his trials. He quickly picked up a piece, tested it for sharpness with his tongue, and cut the rope. It was a moment of great triumph for Kanzi. He had figured out how to make sharp stone fragments.

Kanzi's solution was creative and bonobo-friendly. It was not, as Nick argued, a demonstration that knapping was within an ape's abilities, but a demonstration that there is more than one way to break a rock. Though the shattered pieces Kanzi produced were effective at cutting the cord and obtaining the treat, they were not knapped flakes with conchoidal fractures. They would not be interpreted as tools if they were found in an archeological context.

Attempts to make Kanzi practice knapping again by carpeting the floor were unsuccessful. He simply pulled up the carpet and threw his stones as before. (Bonobos are hugely strong.) The team then moved the experiment to an outside enclosure, where there was bark mulch on the ground. Kanzi's solution was to place a large rock carefully on the mulch and throw another rock at it, producing a satisfactory scatter of shattered pieces. He knew how to break rocks into sharp pieces and he could find a way to do it under trying circumstances. Finally, the researchers found the key to getting Kanzi to knap: they put the rocks into Kanzi's wading pool. Throwing stones into water didn't shatter them. This setup forced him to return his attention to learning how to flake, rather than throwing stones to produce sharp edges. He became a better knapper and is now very adept at it (Figure 16).

In the meantime, Kanzi's younger half sister Panbanisha had been watching these experiments with interest. As Kanzi had, Panbanisha tried a clapping motion that didn't break the rocks or produce flakes, and so, as Kanzi had, she gave up trying to knap. For nearly a year, the experimenters encouraged Panbanisha by making flakes and handing them to her so she could cut the rope that held the box closed, but she wouldn't try to learn to flake herself. Then one day Panbanisha observed Kathy flaking stone instead of Nick. Seeing what Sue calls "an important outside female visitor . . . [and] an expert knapper" at work made a big impression on Panbanisha. She started practicing knapping again, this time paying more attention to the precise movements that Kathy used and having more rapid success than Kanzi had had. Soon Panbanisha was a more proficient knapper than Kanzi, who was apparently jealous of her mastery and often

16. The bonobo Kanzi learned how to flake stone to make tools by watching archeologist Nick Toth.

tried to distract her from the task. Panbanisha was smart enough to use the throwing technique in situations where it would be more successful and the knapping technique in others.

The observations made by Sue, Nick, Kathy, and the others working on the project reveal a great deal about the rich interior life of bonobos and their motivations. It is fascinating that the social interactions among researchers and subjects seemed so crucial to the project's success. Could learning to knap be a fundamentally social process? Or is it simply that bonobos are so fundamentally social as a species—and they are *very* social animals—that to learn they must have a social interaction? I wish I knew.

Twenty years later, the bonobo toolmaking project is still continuing. Kanzi and Panbanisha are now both fairly accomplished knappers and two of Panbanisha's sons, Nyota and Nathan, have started flaking stone too. They have learned primarily from watching Kanzi and Panbanisha, not from watching Kathy or Nick.

This remarkable experiment has provided valuable insights.

First, bonobos are perfectly capable of flaking stone if they are shown how to do it. However, like other animals, bonobos are very good at what they are adapted to do and not very good at what they are not adapted to do.

As a horse trainer of mine used to say, "Horses are very good at horse things. What they are not very good at is people things." This sounds like a trivial observation, but it is really a very profound truth about different types of animals and the limits of their adaptations and inclinations.

Although Kanzi, Panbanisha, Nyota, and Nathan eventually learned to knap effectively, it is clear that knapping is something bonobos have rarely, if ever, been moved to do on their own. The problem is not that they are incapable but that they don't want to. Knapping is not a bonobo "thing" except in these very special circumstances.

Second, if nonhumans had invented a means of making stone tools, the bonobo experiment shows us that there is no reason to suppose they would have chosen the technique used by humans. When Kanzi wanted to flake a stone, he invented a bonobo technique. Different animals, even within the Primates, have different anatomies and different motivations. If a second or third species of hominid (in addition to *Homo*) with significantly different anatomy started making tools, they might also have chosen other ways to obtain sharp rocks.

What is also brilliantly clear is that the bonobos do not have any intrinsic interest in making tools. In the wild, reports of bonobos using tools are extremely scarce; chimps make and use tools much more regularly and in more varied ways than bonobos. As Frans de Waal, a bonobo researcher at the Yerkes National Primate Research Center in Atlanta, puts it, "Tool use in wild bonobos seems undeveloped." Yet in captivity, bonobos have become supreme toolmakers and chimps have not—perhaps because chimps are less interested in social interactions with humans than bonobos. Even once Kanzi could knap, he preferred to make a sharp stone fragment for cutting the rope in his way until he was prevented from doing so. He and the other bonobos in the experiment apparently had little desire to make tools and perhaps lacked the observational skills and attention focus that toolmaking early hominids demonstrably had.

Kanzi and his bonobo community at the Great Ape Trust are not given stones except in the context of knapping experiments or demonstrations (presumably for safety reasons), so we can't know if they would make tools if no humans were present or if no treats had been placed in the box. Seemingly, these bonobos do not realize or do not care that their tools could be used for anything other than cutting open the treat box. A human toddler with a forbidden pair of scissors or a knife will try to cut everything in sight, but Kanzi and Panbanisha don't do this. They don't seem particularly interested in cutting things or in finding out all the things they could do with conchoidal flakes. Cutting apparently has a limited use in their minds. ("Minds" may not be the right word to use here, but I know no more applicable term.)

On my visit to the Great Ape Trust, I and several other researchers sat outside an enclosure watching as both Kanzi and Panbanisha made flakes, cut the twine, and extracted their treats. They clearly knew exactly what they were doing and why. On one trial, however, the lid to the box got stuck and would not open even though the twine had been cut. This unexpected event caused some distress. If I may anthropomorphize, the bonobos saw this outcome as unfair, unwarranted. Each bonobo tried to pry the box open with his or her strong fingers. Sue tried to help. Kanzi understood exactly what the problem was and searched for something useful to help get the box open. After surveying the enclosure and the natural objects in it, he deliberately broke off a length of stem from a tall-growing, woody weed and tried to pry the box open with it. His chosen implement was not sturdy enough and broke immediately. Finally, Sue took the bonobos inside, got a strong screwdriver, and came back out with them to pry the box open.

This incident was very revealing. Kanzi and Panbanisha thoroughly understood the sequence of activities: flake stone, cut twine, open box, get treat. They saw, or had learned, the link between one activity (flaking the stone) and its desirable consequence (getting the treat). They also seemed to me to have some sense of surprise or resentment when the routine failed to work as it usually did. And they demonstrably comprehended that tools would let them do things (solve physical problems) that they could not accomplish without tools. They even had a good sense of the type of tool that would solve this problem, although their choice of raw materials was rather limited. But they used their manual strength, not the sharp stone fragments, to obtain a crude lever; they didn't try to cut a large, sturdier piece of wood to use, although to be fair I couldn't see whether there were more suitable pieces of wood in the enclosure.

Learning how to knap may have prompted them to explore making other types of tools, such as crowbars or levers. In time, Kanzi discovered a second use: flaking became a sort of party trick that he performed to get praise and attention from visiting humans. Typically for a bonobo, his "new" use of flaking was for social interaction.

How good were the bonobos at toolmaking? Hoping to answer this question, Nick and Kathy collaborated with Sileshi Semaw, director of the Gona Research Project in Ethiopia, and the bonobos in a new endeavor. They wanted to compare early stone tools, bonobo stone tools, and stone tools made by modern humans.

The lithic assemblages from two Gona sites, EG10 and EG12, were pooled to create one comparative archeological sample. The second sample was created by Kanzi and Panbanisha using a selection of rocks from Gona. The third "human" assemblage was produced by Nick and Kathy, using Gona rocks. Their aim was simply to reduce the cores by about 50 percent and produce serviceable flakes in the process. After studying and measuring forty-two different aspects on each tool in the collections, the three had a new appreciation of what the bonobos were doing and what had gone on at Gona, the earliest stone tool site in the world.

Technique turned out to be critical. One of the peculiarities about the physiology and anatomy of apes is that they are about five times stronger for their weight than humans. Huge strength is not the same as rapid movement and the bonobos did not accelerate their hammerstones to the same impact velocity that humans do. Nick and Kathy, as expert knappers, achieved an impact velocity of 7.12 meters per second. Kanzi and Panbanisha achieved only 3.67 meters per second. The estimated impact velocity achieved by Gona hominids was intermediate, about 5 meters per second.

The bonobos are more than strong enough to produce the same impact velocity as the humans, so why don't they do it? Nick thinks the toolmaking bonobos may be governing themselves, deliberately not using their full strength or speed. An ill-aimed, hard blow that strikes your fingers is more painful than a tentative one. He also speculates that, having spent their whole lives in captivity, these individual bonobos have always been discouraged from any kind of violent activity. Really slamming one rock with another might seem too much like violence to these bonobos.

The weak force exerted by the bonobos meant that they had to strike the edges of the cores repeatedly to remove a single flake, which was not often true of the humans or Gona hominids. The result was that the bonobos' cores had battered edges, as did their flakes. They reduced their cores by only about 30 percent before they stopped. In contrast to the bonobos, the Gona hominids reduced their cores by 63.5 percent. (The humans—by design—reduced theirs by 50 percent.)

However, the bonobos produced thinner flakes and fewer broken flakes than either the Gona hominids or the humans. Thinner flakes and fewer broken flakes suggest a higher level of skill among the bonobos than humans or Gona hominids. Yet the bonobos produced fewer flakes per core (only 5.36) than either the Gona hominids (9.39 flakes per core) or the humans (11.87 flakes per core). In other words, the bonobos were less efficient at using the raw materials than the modern or ancient hominids.

Was this generally poor performance by the bonobos due to a lack of skill or a lack of motivation?

The shallowness of the bonobos' motivation is a serious issue. Although the bonobos liked the treats, enjoyed interacting with humans, and liked being praised for flaking, it seems improbable that they ever would have invented stone knapping on their own. In the wild, bonobos engage in almost no toolmaking. And in this captive setting, their needs for food and other items (toys, clothing when it was cold, bedding) are always met. Producing flaked stone tools was in no way urgent to these bonobos, nor was it linked to survival. If these bonobos had never produced a single flake, life would have continued the same and they would have been as well cared for as always. (Obviously, allowing the bonobos to suffer if they did not knap was out of the question.)

When these observations of captive bonobos are combined with the thousands of hours of observation of chimps and bonobos in the wild, two things become apparent. First, although wild chimps make and use tools more than any other nonhuman species, they rarely use stone as a raw material for their tools. We don't know how well chimps can knap because no systematic attempt has yet been made to teach them. Second, bonobos are intrinsically not toolmakers. It is not a part of the set of their fundamental adaptations as a species. In fact, chimps and bonobos are neither anatomically adapted to making tools through knapping nor do they seem to perceive a need for sharp implements.

KANZI AND PANBANISHA taught me a lot about making stone tools and about apes.

Yes, these bonobos understand the usefulness of modified objects in obtaining something they want. But their uses of tools and the tools they create are very rudimentary. Bonobos work stone to produce a sharp-edged fragment. They have not diversified their tool types into, say, fine cutting implements, heavy-duty cutting implements, hole-making or piercing tools, scraping tools, or pounding and grinding tools. They haven't explored the possibilities of flaking as a means of making tools, either.

And yes, bonobos and probably chimps are entirely capable of flaking stone, but they don't do so in the wild.

What a contrast with early (and modern) hominids, who not only flake stone but do it over and over! As anthropologists Iain Davidson of the University of New England and Bill McGrew of Cambridge University say of the archeological record, "There seems to have been much more knapping than was necessary to acquire sharp flakes. There does not seem to be any good reason for it, but this excess is quite common throughout the history of stone-knapping (up to and including modern experimental or hobby knappers!)."

Becoming a toolmaker and user was not simply a matter of physical capability for early hominids. Anatomically, apes are quite able to make tools, even those which require skills that are difficult to master. The difference between hominids and apes is that the latter don't seem to feel the making of sharp cutting tools is particularly important. The difference between the impact of toolmaking on apes and on the hominid lineage comes down to five main points:

- First, hominids make tools that transform the raw materials and apes do not. This means that hominids were consciously or unconsciously altering the world in which they lived. In a sense, toolmaking hominids were constructing a new ecological niche for themselves.

- Second, hominids used tools to process dead animals efficiently and swiftly, but apes do not use tools for these purposes. The ability to make tools did not move apes into a predominantly predatory niche and did not provide any pressure to increase their focusing or observational abilities.

- Third, because apes do not use tools to obtain or process prey, toolmaking in itself didn't alter the chimp or bonobo niche.

- Fourth, when apes use tools to obtain food that is otherwise unavailable (like hard-shelled nuts), the new food does not lead to brain expansion, gut reduction, or massive territorial expansion.

- Fifth, by making and using tools as they did, hominids created a situation in which they could reap a substantial selective advantage by focusing their attention closely on both potential prey and potential predatory competitors. Selection for enhanced observational abilities—for being able to watch and understand the activities of other species of animals—may have been very strong. Chimp or bonobo tools do not have this effect, probably because ape tools do not radically increase hunting success and do not increase interference competition.

7

A Brief Stop in the Levant

THE ANIMAL CONNECTION had to become an integral part of the early hominid adaptation as soon as tools were invented and used in the way our ancestors did. Once *Homo erectus* dispersed out of Africa, early humans continued to expand their abilities.

But what were those abilities, exactly? Answering such difficult and particular questions depends in part on the right researcher's discovering the right site—one with remarkable preservation and rich remains.

An Acheulian site known as Gesher Benot Ya'akov in Israel gives us intriguing glimpses of how humans had changed by 790,000 years ago. The site has been excavated for several years by a team led by Naama Goren-Inbar of the Hebrew University. She and her colleagues have found compelling evidence of two important innovations made by *Homo erectus*. The preservation of this site is excellent because it lay on the shores of an ancient lake, so the materials within the site became waterlogged.

The first innovation is the earliest convincing evidence of the controlled use of fire. The importance of controlled fire derives from the fact that heating materially changes the properties of many substances, including many types of food. Cooked foods are more readily digestible (to modern humans) than raw foods because heating breaks down starches, gelatinizes proteins, and generally makes the food item softer. Cooking also kills bacteria adhering to foods.

Food is not the only important resource that is improved by heating. Flint heated to more than about 350 degrees Fahrenheit becomes easier to knap and less likely to fracture across its natural grain. Heat-treated flint can be recognized by changes in luster or color and, more reliably, by a characteristic type of damage termed *potlid fracturing*, in which a small circular piece of stone spalls off of the surface as a result of heating, leaving behind a saucer-shaped depression.

But how can you tell if flint was deliberately heat-treated or not? If flint chips fall into a fire as a tool is being worked, or by accident, the flint will begin showing visible changes at temperatures of 350–500 degrees F.—temperatures that can be easily generated by controlled campfires. Accidental heating might produce a small number of burned spalls or chips, but probably not four burned artifacts and over six hundred burned microartifacts as at Gesher Benot Ya'aqov.

Thermoluminescence (TL) dating was used to test this conclusion further. Thermoluminescence provides a measure of the radiation absorbed in the past by an object; heating above 400 degrees F. resets the thermoluminescent signal of the object, so the signal indicates when a piece of flint was heated, not when the rock formed. Testing nine samples from Gesher Benot Ya'aqov revealed that, as suspected, the eight with evidence of potlid fractures had been heated and the one without such fractures had not. Although burned specimens of flint, wood, and plant remains were scattered through the excavated areas, there were two strong concentrations of burned remains that are interpreted as hearths (Figure 17).

Controlled campfires produce heat of 400 degrees or more; so do wildfires. As the team notes, although there are substantial numbers of specimens that have been heated (judging from their appearance and TL signatures), the burned items amount to less that 2 percent of the recovered flints and plant fragments. Further, these burned items are spatially clustered into two areas (interpreted as hearths). Wildfire could have produced the charring of the wood and the fracturing of the flint, but not the concentration of burned specimens.

Added to the evidence of heating, the spatial distribution of the burned materials enabled Goren-Inbar to conclude that these were controlled fires or hearths. Lightning strikes are an unlikely cause of the burning seen at Gesher Benot Ya'akov because they start grass fires, which create a widespread distribution of burned remains, not tight concentrations. Only 2 percent of the flint and wood remains are burned—much less than would be expected from a widespread fire— and there are quantities of unburned driftwood in the sediments at the site. The driftwood should have burned too if there had been a widespread fire. Finally, the remains of mollusks, crabs, fish, and mammals suggest the site was formed during the rainy season when few spontaneous fires occur. Thus the evidence for controlled fire is strong at Gesher Benot Ya'akov, and no earlier sites offer such compelling evidence.

In Level 2 of the excavation, the evidence is particularly striking. There were spatially distinct areas where different raw

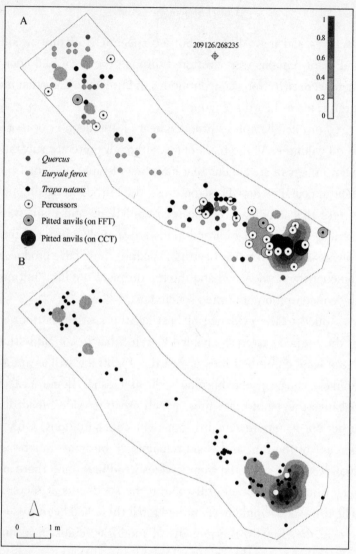

17. *The remains excavated from Level 2 at Gesher Benot Ya'aqov show distinct clusters of items associated with particular tasks. (A) Nut-cracking areas are shown by the locations of pitted anvils; different kinds of edible seeds (Quercus, Euryale, and Trapa); and percussors, which are superimposed on a density map of burned flint microartifacts. The two clusters of microartifacts are interpreted as being hearths. (B) The locations of seventy-four wood remains (black dots) are not tightly clustered with the hearths, except for the burned wood specimens (white circles).*

materials were knapped: flint mostly in the northwest of the site, while basalt and limestone (as well as a few heated flint specimens) were concentrated around the hearth in the southeast. In other words, the hominids didn't make tools anywhere they felt like it, but only in particular areas. Tools knapped on both sides—bifaces—were modified for various tasks, for example, near the hearth. Fifty-four artifacts from the site, including anvils and hammerstones, have pitted surfaces from being used to break open nuts or kernels (Figure 18), and these are also clustered. Particular food remains—nuts, fish, and crabs—were processed in the same area and may have been intentionally cooked or heated.

In addition to the evidence of fire, the outstanding preservation of plant materials at Gesher Benot Ya'aqov yields precise information about the plants used by the humans who inhabited the site. Remains of nut shells that fit neatly into some of the pits on the anvils were found at the same level. In total, Goren-Inbar's team has excavated plant remains from seven different species that yield edible fruits or nuts: wild almond, prickly waterlily, two types of pistachio, water chestnuts, oaks, and an evergreen shrub. None of the fruits of these species can be opened by hand, so the association of pitted stones and plant remains suggests intensive use of this resource. Further, the nuts of some of these species require roasting to reduce or neutralize toxins before they can be readily eaten. Human knowledge of plant remains and the edible resources they offered was impressive.

The excavation and preservation of remains at Gesher Benot Ya'aqov reveal a great deal about *Homo erectus*. These hominids clearly partitioned their space up into activity areas, a behavior

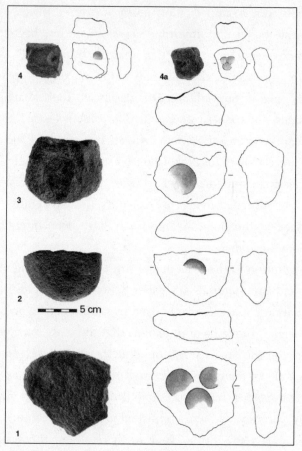

18. *These specimens from Gesher Benot Ya'aqov, labeled 1–4, show the progression of wear on nut-cracking anvils made of basalt. No. 1 shows incipient pits; No. 2 shows a shallow pit; No. 3 shows a large, deep pit; and No. 4 shows a series of pits well developed from repeated use.*

that is often considered modern and sophisticated in some ways. Setting aside different areas for different activities is seen by the investigators as indicating a "formalized conceptualization of living space," which is usually considered to reflect modern sensibilities. In people today, the use of space may reflect such things as an individual's kinship, his or her age or sex, social status, and knowledge or skill.

Our ancestors were more advanced, even 750,000 years ago, than has been widely assumed. Why have paleoanthropologists misjudged our past in this way? First, 750,000 years ago—or 1–2 million years ago—seems impossibly distant. For some, it is hard to imagine that people who lived so long ago could have been modern. Second, only sites with the very best preservation that are meticulously excavated can possibly reveal such insights. Recent advances in dating and in excavation and analysis techniques have made a profound difference in our view of the ancient world.

As the distribution of humans expanded out of Africa, into the Levant, and on across the Old World, so did their abilities. Tools became more complex and were made of a wider range of materials using some new techniques, including soft hammer percussion in which a bone or antler is used as the hammerstone. More and different kinds of animals were hunted and eaten; sources of food once often left untouched, such as fish and shellfish or toxic nuts, were gathered and processed in complex ways to make them edible.

And one adaptation—one remarkable ability that was predicated upon the need for information about animals—eventually arose in our lineage.

8

Say What?

INVENTING STONE TOOLS forged an adaptive connection between our ancestors and the animals in their ecosystem. Changing the way our ancestors fed and what they needed to know in order to be successful was only the first great step in human evolution. Knowledge became precious and the sharing of knowledge of many kinds became an adaptive advantage. Thus the second great step in human evolution was the invention of language.

Since the time of Aristotle, language has been considered a primary hallmark of humanity almost as often as toolmaking has been. Like tool using, our possession of language has been a source of pride and boasting. Humans often see language as uniquely human and essential to our success in the world, though linguists battle fiercely over their different ideas about how language contributed to that success.

Clearly, full language would have conferred substantial advantages to a species that already had an ability to adapt extrasomatically. Tools—the first great extrasomatic adaptation—

allowed humans to function as if they had certain physical adaptations without bothering to evolve them.

What specific advantages did language offer? Language allowed humans to function as if they had certain types of knowledge without having to acquire them individually.

Over time, the advantages of paying close attention to other animals and learning about them would have racheted up. I believe that having detailed information became more and more crucial to survival and evolutionary success. If I had some knowledge but you had more, you might easily get more food, escape more tigers, or simply live longer to have more offspring. If I knew where to gather nuts rich in protein and fats and how to detoxify them, I had a fallback resource that would be unavailable to anyone without this knowledge. In humans, collecting information about other species and paying close attention helped us survive the perilous transition from being strictly a prey species to being both a prey species and a predator. So, to the extent that it helped us collect information faster, better, and more reliably, language was a huge advantage.

At its essence, language is a primary and external means of storing, ordering, and transmitting information to others. Like making tools, language is another sort of extrasomatic adaptation that allows humans to gain information incredibly rapidly, rather than slowly and laboriously.

What difference would language make? A huge one. Suppose one individual in a group learns from experience about lion-hunting behavior and how it changes in times when there is a lot of game as opposed to times when there is very little game. Simply observing these disparate pieces of knowledge about what lions do, recognizing variation in the behavioral and ecological pattern, and comparing observations from one season to another requires sophisticated processing and organization of information in the brain. Then suppose that some individual develops a mechanism for sharing that digested and synthesized information with another individual though language. Suddenly, information is no longer based on a single individual's life experience but becomes cumulative. Perhaps the second individual has witnessed differences in how aggressive lions are in defending a kill from competitors based on the amount of vegetation and cover at the kill site. This variable, too, can suddenly be taken into account if the information can be shared.

Organizing and communicating information is clearly the primary use of language today. If this has always been the function of language, then the archeological record ought to show significant changes when language arose. Being able to communicate effectively and exchange valuable information with others should have been a quantum leap for our ancestors in terms of their populations and their survival.

Why, then, are there such serious debates over the origin of language and its timing? The disagreements are particularly heated perhaps because there is so little direct evidence and so much theory. Language does not fossilize and words don't leave diagnostic marks on fossilized bones. Language isn't made possible by a tangible organ or bone, so anatomical studies aren't as helpful as they might be. This means that most scholars in the field rely on intricate analyses of modern language as the evidence to support their assertions. And—as with the hackneyed creationist challenge, "what good is half an eye?"—if you use current adaptations and benefits to intuit the evolutionary pathway that led to a structure's current form, you can go far astray.

Before we can search the fossil and archeological record for evidence of language origins, we have to know exactly what we mean by "language." This is a far from trivial exercise.

What is language?

First and foremost, language is an innate, complex capacity of modern humans. Noam Chomsky is a professor at MIT, the founder of modern linguistics, and a towering figure in the field. He has proposed that there is a specialization of the human brain—and only the human brain—that he calls a "language faculty" or "language organ." What he refers to is a cognitive or intellectual ability of the human brain, not a physical entity that you can find in a dissection. Language is not confined to the spoken word but also includes written and organized gestural systems. Thus, an inability to form words because of the shape of the oral cavity or the positioning of the tongue or of the larynx in the throat does not equate to a failure of language. Mute humans do not lack language; they lack *speech*. Humans who have had their larynx removed for medical reasons do not lose their language ability. And although apes cannot form the spoken word as humans do, that fact does not rule out the possibility that they could have language.

Speakers of Ameslan, the predominant deaf language used in the United States, are not using a word-for-word substitution of specific gesture for specific words. Ameslan has grammar and syntax, but they are not the same as those of English.

All normal humans have the ability to learn a language or languages, though there is no sign of an ability or genetic predisposition to learn a particular language.

Whether or not other animals also have language facilities remains controversial. Chomsky once said, "If an animal had a capacity as biologically sophisticated as language but somehow hadn't used it until now, it would be an evolutionary miracle." His assumption—and it is a very large assumption—is that humans would notice, understand, and recognize an animal language if one exists.

Many humans today have little contact with or understanding of animals, witness the foolish people who think they can keep a tiger in a city apartment. There are others less dumb to animal needs who nonetheless buy a dog bred for herding or running and then wonder why the dog is unhappy when its only exercise is a brisk walk around a city block. Similarly, many people with no experience of horses think horseback riding involves "putting the horse in gear" and simply steering. As one who lives with and pays attention to animals, I sometimes feel part of a dwindling group. I would not be entirely surprised if a language among wild animals went generally unnoticed. Most of us are rarely exposed to animals, and still more rarely do we pay the close attention to animals that would be needed to detect a language if animals were using one.

Language is also an intrinsically symbolic behavior. Words (or icons or signs) are not usually representational; they are symbolic and arbitrary. Words can and do refer to imaginary items (like unicorns), tangible items (like shoes or giraffes), intangible items (like emotions or colors), and actions (like jumping) in ways that are not necessarily a direct reflection of their content. In order for a symbol to function in communication, it must also be repeated, so that the symbol and the concept become identified with each other and interchangeable. And, of course, the meaning of the symbol must be shared or recognized by more than one individual.

According to linguists like Steven Pinker of Harvard University and his colleague Ray Jackendoff of Tufts University, language is fundamentally communicative. They say that "the language faculty evolved in the human lineage for the communication of complex propositions." Similarly, anthropologists William Noble and Iain Davidson, both of the University of New England in Australia, define language as "the symbolic use of communicative signs; the use of signs in communicative settings to engage in acts of reference." No other animals use signs and symbols in this way. Although ape language experiments show that apes can use signs and symbols, they don't do so in the wild as far as we know.

Certainly, language is also used to organize thought, which Chomsky emphasizes over communication. Case studies concerning individuals who had full use of language and then experienced a brain injury—a lesion or a small stroke, say—are very revealing. Such patients often suffer a partial loss of language or aphasia as a result of their injury. The extent of loss is directly related to the extent of the injury: the less brain tissue that is damaged, the milder the aphasia. In very small, discrete injuries, the loss of language reveals how precisely concepts or words are stored in the brain. For example, injury to one small area may result in the loss of proper nouns; in another, the loss of verbs. Amazingly specific deficits have been reported for color words, body parts, household objects, and for fruits and vegetables.

A student of mine once told me about his grandfather, who had been a career track coach. He suffered a small stroke; afterwards, he was unable to produce any nouns related to running, such as shoes, hurdles, sprint, and the like. He was deeply frustrated to have lost fluency about a topic that had been so central to his life. He did not lose the understanding or experience he had gained as a coach, but he had lost the nouns through which he stored that knowledge. In his brain, these nouns were apparently stored together in immediately adjacent areas that had been damaged.

Although language is a means by which we store information for our own later retrieval, is it primarily about communication? I think so. My student's grandfather had not forgotten what he knew about running but was merely unable to summon the words to talk with others about it any longer.

Language is also intrinsically social. The acquisition of language requires a minimum of two individuals, if not a community. The fact is that socially isolated individuals do not develop a language, not even for speaking to themselves.

The textbook example is a girl named Genie (a pseudonym), who was horribly abused and is probably the worst case of social isolation studied in modern times. She was tied to a potty chair, locked in a room, and isolated from meaningful human contact from the age of about sixteen months until she was discovered in Los Angeles at thirteen, in 1970. She had apparently been beaten for making any sounds during her confinement. Other than that, she had had minimal social contact. Whether or not she was mentally retarded before this abusive treatment is unclear, but when she was rescued, Genie did not speak or understand speech. Not surprisingly, she had numerous other physical, mental, and social difficulties. Despite years of intensive work to rehabilitate her, Genie never acquired full language abilities.

When she began speaking, her utterances were very simple: "Mike paint." "Father take piece wood. Hit. Cry." At her best, after four years of tutoring, Genie could pronounce strings of up to six or seven words that might be considered ungrammatical sentences, such as, "Teacher said Genie have temper tantrum outside." After a series of disputes and lawsuits over her care and the research, Genie was taken from the family home of one of the team of researchers, where she had been living for years, and was returned to her mother, who had been acquitted of abusing Genie. When the mother found herself unable to provide the nonstop care Genie required, the child was sent into foster care, where she was again abused and mistreated. Now an adult, Genie has reverted to silence and lives in a home for the retarded.

The study of Genie was one of the primary cases that supported the idea of a critical window of opportunity for learning language which, in her tragic case, was lost. Genie did not learn language because there was no one to speak to and no one spoke to her at the right period in her life. Exactly when that critical period is during a child's lifetime is not clear. Some linguists maintain the critical period is between birth and age six; others say simply that if a child is exposed to normal language before puberty, there is reasonable hope for normal language acquisition.

Genie certainly failed to meet the criteria of acquiring full language. Her utterances frequently have clear meanings, but— as researcher Susan Curtiss says in a television documentary on Genie—her utterances are not sentences in English. What is also crystal clear from the videos of Genie is that she was eager to learn the names for things and to have new experiences. Curtiss recounts an incident in which Genie demanded to know the name for every color of thread in a Wal-Mart display. Her

deep curiosity and intelligence and her enthusiasm for learning are unmistakable. Moreover, Genie's drive to learn language
and new words exceeds by an order of magnitude the interest
in language and communication shown by apes in the language
experiments that I will talk about in chapter 11.

There is another telling case, an illegitimate girl named
Isabelle who was confined from birth until she was six years
old with her deaf-mute and retarded mother. Isabelle was not
socially isolated to the extreme extent that Genie was and was
discovered at a much earlier age. She heard no speech until age
six, but she and her mother communicated crudely through
gestures, indicating that the idea of communicating to another
person had taken hold in Isabelle's mind. After two years of
intensive therapy, and medical treatment for the rickets she
acquired because she was never allowed outside, Isabelle had a
vocabulary of 1,500–2,000 words, a normal command of syntax,
and was able to function normally at school with other children
her age. One of her documented utterances (after training) was:
"Why does the paste come out if one upsets the jar?" This is a
much more sophisticated and complex utterance than anything
Genie or any trained ape ever said. Perhaps because Isabelle's
social isolation was nowhere near as severe or long-lasting as
Genie's, Isabelle was able to master language after her discovery.

Tim Ingold of the University of Aberdeen argues that language is not simply a social ability but arose primarily in order to
exchange social or interpersonal information. He claims that the
original function of language was to identify people, to recognize
the self or us as opposed to them. Similarly, primatologist Robin
Dunbar of Oxford University has hypothesized that gossip was
the primary engine driving the evolution of language. In his

view, exchanging information about social relationships through gossip is a form of verbal grooming, a means to ensure that the social bonds among individuals in a group are maintained.

In the next chapter, I explain why I disagree with Ingold and Dunbar about the original function of language. Language's primary use today may be to facilitate social interactions and bonding, but there is evidence that language, at its origin, had another primary function.

The universality of language among normal humans, and its symbolic, social, communicative, and organizational aspects, are reasonably clearcut and agreed upon by many who study the origin of language.

However, there are additional aspects of modern language that may or may not have been intrinsic to language from the time of its origin. Modern or full language includes rules for arranging those symbols to convey meaning, which is called syntax or grammar. "Woman kills deer" does not mean the same thing as "Deer kills woman," despite the limited number of words involved and the similar grammatical structures.

Chomsky is the foremost advocate of the proposition that language is fundamentally syntactical and that forms of communication lacking syntax are not language. His opinion derives from watching how children learn language.

A human language is a system of remarkable complexity. To come to know a human language would be an extraordinary intellectual achievement for a creature not specifically designed to accomplish this task. A normal child acquires this knowledge on relatively slight exposure and without specific training. He can then quite effortlessly make use

of an intricate structure of specific rules and guiding principles to convey his thoughts and feelings to others, arousing
in them novel ideas and subtle perceptions and judgments.

In other words, Chomsky believes no child actually learns syntax and grammatical structure; those essentials are innate in the
human brain. The child only learns the local (regional) vocabulary and particulars to fill out the inherent syntax encoded in the
infant brain.

Chomsky has long defined syntax as *the* essential component
of language and he now specifies a particular element of syntax,
termed *recursion*, as key. Embedded or recursive elements are
sentences or clauses within sentences. "I made a handax out of
basalt" is a straightforward sentence with no recursive elements.
"You know that I made a handax out of basalt" contains a recursive element ("I made a handax out of basalt") which is the same
as the entire sentence in the first utterance. Recursive elements
can go on seemingly infinitely. Simply add the words "Kathy
knows that" to the beginning of each of the sentences above,
and you will see what I mean. The ability to embed clauses or
sentences within other sentences—and convey meaning—is
what permits an infinite number of meaningful sentences to be
constructed.

If syntax is essential to language, then we must accept that
when children begin to speak (or sign), they are speaking a nonlanguage that has nothing to do with the real language which
they will acquire by about age six, through some independent
mechanism. Do small children speak a nonlanguage? Personally,
I don't think so. Although conversations with my grandchildren
when they were very young occasionally made me feel as if we

lived on different planets, I had no doubt that they were using and misusing the same language I was in the typical toddler (ungrammatical and asyntactical) fashion.

Full modern language also involves special grammatical items that enhance communication and the transmission of meaning. These can be termed *disambiguators*, the items that permit humans to express complex relational or referential ideas and to speak of the past, the present, the future, and the hypothetical.

Let's return to another one of Genie's utterances—"Tissue paper blue rub face. Pound."—to see what disambiguators do. According to Susan Curtiss (who was teaching Genie at the time), this utterance meant, when disambiguated, "Father used to rub my face hard with blue tissue paper." What "Pound" meant is not clear. Alternatively, the utterance might be construed to mean "If I rub my face with blue tissue paper, I will get pounded (beaten)." Or, "I rub my face with blue tissue paper and then pound (flatten) it." The additional words needed to distinguish these possible meanings from each other and from alternatives are disambiguators.

Because pidgin speakers also use few disambiguators, very simplified vocabularies, and loose syntax, Derek Bickerton of the University of Hawaii has proposed that pidgin languages can serve as an informative model of protolanguage, or the type of communication that must have preceded full language. Pidgin is probably not a perfect model of a protolanguage, since pidgin languages are invented and used by modern humans who already have language; but pidgin shows what modern humans think is essential to language. The sentences or strings of words are typically short, and the order of verbs, adjectives, nouns, and parts of speech is variable. Verbs do not necessarily agree with subjects and, in some cases, verbs are omitted entirely. Bickerton cites examples of remarks made by immigrants and plantation workers in Hawaii between 1880 and 1930.

One is: "Aena to macha churen, samawl churen, haus mani pei." Written in English, this reads as: "And too much children, small children, house money pay." The translation Bickerton offers is: "And I had many children, small children, and I had to pay the rent."

Indeed, Bickerton's studies of pidgin show that achieving the function of language—the communication of complex information—does not depend upon all of the constituents of modern language being present. If all aspects of modern language—syntax, grammar, tenses, recursion, and disambiguators—are necessary for communication to qualify as language, then most probably language arose as a unified genetic package: a single mutation or a small group of linked mutations. There was the time before language and then—boom!—there was language, in one generation.

While the idea of a language miracle is appealing in its simplicity, I cannot help but worry about the first possessor of this mutation and with whom he or she talked. And if no one else nearby had the same mutation, what then?

Do the various elements of language come as a whole, unified, and indissoluble package? No. Numerous scholars who know more about linguistics than I do object to Chomsky's proposition that syntax is an essential and integral component of language. When Genie said, "Teacher said Genie have temper tantrum outside" (which I translate as, "Teacher said that Genie was to have her temper tantrum outside"), Genie was not using full language as I would recognize it, but she was using a recursive structure. Similarly, linguistically trained apes that I shall discuss in chapter 11 occasionally use recursion, but they don't have a full command of language in Chomsky's terms either. From such evidence, I conclude that the elements of modern language are not intrinsically linked and were not all present at the beginning. Some elements—probably including syntax– could have evolved well after the initial invention of language.

Deducing the sequence in which the elements of modern language evolved is a fascinating and difficult task. There is superb evidence that comprehension of words, production of words, and the grammatical arrangements of words form an essential sequence, at least in modern humans. From this work, we get tantalizing hints of how language might have arisen.

The work carried out in the late 1990s by Elizabeth Bates of the University of California at San Diego and Judith C. Goodman of the University of Missouri, Columbia, on language acquisition provides important insights when coupled with work by other researchers.

Most parents and researchers understand the typical sequence of language development in normal children. They begin babbling at about 3–4 months of age and move to using combinations of vowels and consonants at 6–12 months. Meaningful speech begins on average around 10–12 months, probably after the ability to comprehend speech appears. Vocabulary grows fairly slowly but usually undergoes a clear acceleration at 16–20 months. Children start to use combinations of words between 18 and 20 months, and another leap in linguistic ability often occurs between 24 and 30 months. By 36–42 months of age, nearly all normal children have acquired the basics of language.

But how are the different aspects of speech related to each other? Bates and Goodman studied word comprehension, word production, and grammar, starting with children at 8 months of age and continuing until they were 30 months old. Comprehension, production, and grammar increase similarly with age, but they don't progress simultaneously. Among the youngest children in the study (8-month infants), word comprehension was low, about 5 percent of test words. Word production in children did not reach the 5 percent level until later, at about 14 months. The same level of grammatical utterances came still later, at about 21 months. In other words, understanding preceded production and production preceded complex structure (grammar). This is a reassuringly sensible outcome that mirrors the way adults learn foreign languages—or the way adults create pidgin languages, which never progress to the level of complex structure.

In a second study, the Bates-Goodman team focused on an even more interesting phenomenon. They wondered how tightly the acquisition of vocabulary was linked to the development of grammar. Was there a crucial threshold of vocabulary size that had to be met before grammatical construction occurred? Yes.

The single best predictor of grammatical ability at 28 months is a child's total vocabulary size at 20 months—a finding that was highly significant statistically. When a child comprehends about 200 words, expressive language takes off. This is the first threshold in the mastery of language.

The next threshold involves grammatical complexity, a further step in self-expression. Although grammar and vocabulary are tightly coupled between the ages of 16 and 30 months, it is not age (or maturity) in itself that controls grammatical mastery. Instead, the sheer size of the child's vocabulary is the best predictor of the complexity of a child's grammar. Once a child is able to produce about 400 words—the second threshold—grammatical complexity soars.

The relationship between vocabulary size and grammatical complexity is so robust that it holds true even in children at both ends of the language spectrum. Early and late talkers show the same relationship between comprehension, vocabulary size, and grammar. Children with Down syndrome have delayed language acquisition but also show the link between vocabulary size and grammar. So do individuals with Williams syndrome, who have low IQs but remarkable linguistic fluency. Across this full range of language abilities, humans with fewer than 400 words are apparently not able to make utterances of any grammatical complexity and their utterances tend to be very short.

If language evolved from some other form of communication, then would this same pattern be evident in other forms of communication or not? To answer this question, we have to think carefully about animal communication and compare it with human communication.

All communication (including language) shares three components:

First, there is an intended audience. In normal form, communication involves at a minimum two individuals and a transfer of content between them. Communication could also mean documenting or storing a piece of information as an aide-mémoire intended for yourself at a later time.

Second, there is a symbolic vocabulary shared by both communicator and the intended audience. Without this, content cannot be transferred.

Third, there is a specific content or subject to each communication.

The three components of communication listed above are found in bird calls, for example, or alarm calls emitted by meerkats, as well as in signed, spoken, or written communications among humans. But full language goes beyond these fundamental aspects of communication to provide a richness and generative capacity—the possibility of creating endless novel sentences—that is not shared by communication.

It could be argued that the call given by vervet monkeys to indicate imminent danger from a flying predator like an eagle is also symbolic. It is not an enactment of a hawk swooping down or a mimicking of the eagle's cry. The call vervets give does not even mean "eagle." The best translation I can propose of this call is "dangereagle." That is, both an assessment of the situation—danger—and the general type of danger—from a flying predator—are combined into a single utterance. "Dangereagle" cannot be used to inquire about whether or not there was an eagle in this place yesterday, nor can it be used to find out if the eagle that caught a fellow vervet had barred wings or a crest. "Dangereagle" is not specific enough to be used in such conversations. It encompasses both the concept of immediate danger and the concept of eagle, as if they were one thing. However, "dangereagle" is different from "dangersnake" and "dangerleopard," and vervets clearly understand the distinctions.

Certainly, "dangereagle" is communication. It is a cry intended to be understood by another individual, it involves an arbitrary symbol (the cry) with a shared meaning (other vervet monkeys react to the cry); and there is patently specific content, since other cries used by vervets do not have the same meaning or produce the same reaction.

Animal communications in the wild do not observably involve syntax, perhaps since they rarely involve strings of symbols (words). The cry I have translated as "dangereagle" could equally be translated "eagledanger," "flyingdanger," or "dangerup." All of these hypothetical translations omit verbs or action words, except for the implied immediate presence of the dangereagle. There is no past tense, no future tense, and no talking about imaginary, purple-spotted, monkey-eating flyingdangers.

Scholars who have written about the origin of language usually point to such animal communications and note that the calls or cries that have specific meanings are not able to be combined with each other. In *Adam's Tongue*, Derek Bickerton scoffs at the idea that animal communication systems, like that of the vervet monkey, could evolve into true language. "It's not that animals are too dumb to put things together," he says. "Just that the calls and signs and all the other things they communicate with weren't designed to be put together. . . . 'Dangerous food'? Unlikely; danger calls, as we've seen, at least roughly specify the source of the danger without further addition. . . . 'Edible danger'? Come on!"

As symbols or words, these animal cries say too much, rather than just enough. They describe an entire situation, not its components in neatly separable—and recombinable—units. Thus Bickerton, like some other linguists, does not believe that language arose from animal communication systems, but independently.

I'm not wholly convinced by Bickerton's argument on this

point. As the work by Bates and Goodman suggests, possibly animal communication simply does not involve large enough vocabularies. If vervets had more "words," mightn't they have "dangereagle" plus a more precise word that refers simply to a large predatory bird that is not currently present? Perhaps animal communication does tell us how language started but reveals that animals other than humans simply don't have enough words.

Nonetheless, this classic example of animal communication highlights the survival value of information and communication and pinpoints some differences between animal communication and human language. Being able to warn other members of the group (kin) has a selective advantage for vervet monkeys, but apparently more extensive discussions about eagles or snakes does not. For humans, having much more detailed information about potential prey or competitors (beyond their mere presence) was sufficiently crucial to impart a selective advantage to those who paid most attention to other species and to those who shared their information. Enhanced or more specific and complex communication also obviously carried a selective advantage, especially to hominids who had changed their position in the ecosystem in which they lived. Two anthropologists, John Tooby of the University of California, Santa Barbara, and Irven De Vore of Harvard University, use the term *informavore* to express how essential the need for information is and was to hominids, as if information were literally a food to be consumed.

To have information—to collect it by observing the world—was valuable from the onset of the animal connection. To transmit information and share it with others is a more difficult but more valuable task that requires a tighter social group, more cognitive abilities, and a more expansive use of symbols. Let us define language pragmatically as the mechanism—or tool—humans evolved to perform that task.

The problem is determining when language arose in evolutionary history. How can we possibly measure or assess the cognitive abilities of extinct species or past peoples? In truth, we can't, not reliably.

9

Tell Me All About It

In 1991, William Noble and Iain Davidson reviewed the problems in defining and recognizing language in the archeological and fossil record. They realized that language in itself wasn't likely to be visible in the archeological record until the advent of written languages—a far too recent event to coincide with the origin of language itself. They hit upon the idea of looking not for language but for one of its components: symbols. They proposed an admirably commonsense solution to identifying the origin of language in the archeological record: "An approach is needed that is founded in the special nature of language, which differs from the communication systems of other animals in that its constituent signs are used symbolically. The point of origin of language was thus the point at which the symbolic property of signs was first discovered."

Simple, isn't it? Because you can't see words in the fossil or archeological record, use other symbolic behaviors as a proxy for language. At the heart of Noble and Davidson's approach was a sort of checklist of abilities that they and others felt demonstrated intelligence sufficient for language, and thus for modern (human) behavior. Archeologists have been putting together sets of advanced behaviors like this for decades. Depending on the scholar, these sets of modern or advanced behaviors might include being anatomically modern, planning ahead, organizing space into discrete task areas, using symbols, making art, trading goods with other groups, hunting a broad range of animals, fish, and birds, making more sophisticated tools out of a variety of raw materials, or creating items of personal adornment. The idea is that these abilities were all somehow linked by a certain level of cognition or intellectual ability. Therefore, all these abilities should come together, as a package. If you had one of these abilities, then you should have them all in short order.

We do know when humans first became anatomically modern—that is, looked and could have presumably functioned as we do. Anatomical modernity is not necessarily the same as behavioral modernity and many anthropologists have speculated about the "time lag" between the appearance of the oldest modern human and the oldest modern human behavior. The oldest recognizably modern skulls were found in the Omo Kibish Formation in Ethiopia, which was recently securely dated to nearly 200,000 years ago.

If modern anatomy indicates only the potential for but not necessarily the practice of modern human behavior, what should we be looking for? Trait lists of modern human behaviors that were part of the "human revolution" were originally constructed based on the European archeological record as it was known early in the twentieth century. In Europe, the attributes on the list seem to turn up abruptly and nearly simultaneously, at about 50,000–40,000 years ago. Most paleoanthropologists concurred that the "human revolution"—the appearance of modern human behavior—occurred then.

Then, a few years ago, Sally McBrearty of the University of Connecticut and Alison Brooks of George Washington University undertook a major review of the African evidence, which led them to argue fervently that these criteria or trait lists for behavioral modernity are flatly misleading. In their view, the so-called "human revolution" or "Upper Paleolithic Revolution" that so many scholars have said occurred at about 40,000–50,000 years ago and ushered in modern human behaviors was not a revolution and simply didn't happen. The paper written by McBrearty and Brooks was entitled "The Revolution That Wasn't," and it caused a great stir and a major rethinking in the field.

In 2003, Christopher Henshilwood of the University of Witwatersrand and the University of Bergen and Curtis Marean, of Arizona State University's Institute for Human Origins, suggested that "the key criterion for modern human behavior is not the capacity for symbolic thought but the use of symbolism to organize behavior."

Fine, but what does "symbolic organization"—perhaps art, ritual, or music—look like in the fossil and archeological record?

Evidence for ritual is ancient but not entirely unequivocal. At about 1.5 million years ago, a *Homo habilis* skull from Sterkfontein in South Africa acquired cutmarks on its cheekbones and inside its bony orbits. Though the location of the marks is similar to those created by defleshing—taking the meat off—animal skulls, the marks don't prove this individual was eaten by another hominid. There is no indication what was done with the flesh after it was removed from the skull. Defleshing might have been carried out for ritual purposes (as might cannibalism). If the dead were left where they fell and the smell of decaying flesh was offensive, taking the flesh off a rotting bone might have been no more than ancient "housekeeping."

In another example, 600,000 years ago at Bodo, Ethiopia, someone defleshed an archaic (not modern) male human skull, again leaving cutmarks. More compelling evidence comes from the three skulls from Herto Bouri, Ethiopia. This is the same general area in which 2.5 million-year-old bones with cutmarks have been found, but the skulls are from a much more recent geological layer that is only 154,000 years old. The Herto Bouri skulls are the oldest largely complete modern human skulls yet found. All of the Herto Bouri skulls bear cutmarks, and the single child's skull was both cut-marked and carried around for long enough to acquire polish from handling (Figure 19). Why would someone clean and carry around a child's skull? We can only imagine. While cutmarks and curation—archeologists' jargon for "carrying objects around"—do not prove very ancient ritual beliefs, they strongly suggest ritual behavior.

If these earliest examples of ritual treatment of skulls are discarded as unproven, there are unmistakable burials of anatomically modern humans with red ochre and grave goods dating to at least 130,000 years ago and Neandertal burials starting at about 70,000 years ago. Those surely indicate ritual treatment of human remains and imply a belief in some sort of afterlife.

What about art? The oldest artworks in Europe are the exquisite, tiny, carved animals and birds that have been dated to between 32,000 and 36,000 years ago from Vogelherd in southwestern Germany (Figure 20), and cave paintings at Chauvet in the Ardeche region of France dated to the same age.

Compared with evidence from the African record, Euro-

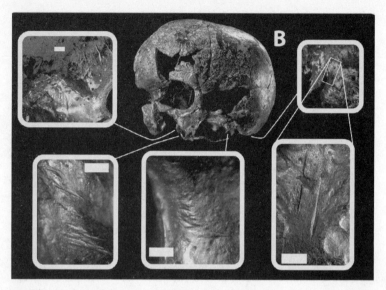

19. *The cranium of the anatomically modern child from Herto Bouri in Ethiopia shows a series of cutmarks (insets) produced during defleshing. There is also polish and wear suggesting that the cranium was carried around for some time, probably hanging from a hide thong. This is some of the earliest evidence for ritual behavior by our early ancestors. The scale bar in the photos is 1 mm long.*

pean art is very recent indeed. Ochre was mined over 300,000 years ago at Twin Rivers, Zambia; many ochre pieces were found at Kapthurin, Kenya, a site 282,000 years old. Was that the beginning of symbolism? Maybe. The commonest use of ochre is as pigment, to paint skin, hides, beads, pottery, house walls, or in art. The difficulty is that ochre also has some medicinal value and can be used in tanning hides, which uses are not obviously symbolic behaviors. Are the lumps of ochre at Twin Rivers and Kapthurin evidence of symbolism? We don't know—cannot prove—what the ochre was used for. Anthropologists line up on both sides of the debate.

Direct evidence of symbolic ochre use—dating to between 100,000 and 77,000 years ago—has recently been reported from Blombos Cave. Blombos Cave is a remarkable site in Western Cape Province, South Africa, being investigated by a team led by Christopher Henshilwood. So far, the cave has yielded fifteen pieces of ochre that are unmistakably incised. The scratched designs have been examined microscopically by Francesco d'Errico of the University of Bordeaux, who also worked on bone tools with Lucinda Backwell. Over the years, Francesco and I have worked on such parallel topics that, despite meeting rarely, we have become friends. Though the significance or meanings of the scratches is elusive—at least to me—they are clearly artificial, intentional (as opposed to accidental), and humanly made (Figure 21).

Francesco says that the scratches "are not consistent with

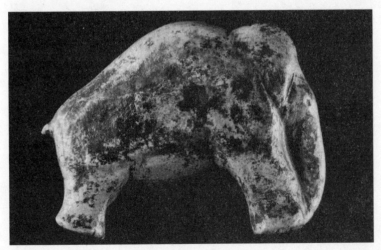

20. *This exquisite mammoth sculpture was carved from mammoth ivory over 35,000 years ago. It is only about 11/2 inches long and was excavated at the site of Vogelherd in Germany. It is one of the earliest art objects in Europe.*

doodling, but rather represent a focused and not abstracted attempt to produce a pattern. These markings are also not notations," he goes on to explain. "A 'notation' can be defined as a marking system specifically conceived to record, store, and recover information outside the physical body. . . ." The problem is, the marks on the Blombos ochre join each other or overlie each other; it is difficult to decide what exactly a discrete marking *is* on these specimens.

Francesco describes the marks in more detail, saying, "The most striking examples from Blombos are geometrical designs that might represent basketry, weaving, fences, or any number of more abstract concepts. Because they share so many features

21. *The ochre piece in the middle ground is one of nineteen specimens from Blombos Cave, South Africa, showing geometric carvings—a form of art. The ochre pieces are between 75,000 and 100,000 years old. Also from Blombos are finely tapered bone points for spears (foreground) and sophisticated flaked stone tools (background and upper right). The paintbrush in the foreground is for scale.*

in common and were made over a significant period of time, these engravings qualify as a *tradition*" (Italics in the original).

Frustrating as it is, we can see these engravings are meaningful—literally full of meaning—but we cannot understand their meaning. We do not share the symbolic vocabulary. We can only know that one existed.

Remarkably, a recent inventory of geometric or nonfigural art in the cave art of France, which is much younger (only 35,000–10,000 years old), shows some striking similarities to the Blombos Cave markings, as do later occurrences in other regions of Africa, the Americas, Asia, and Australia. Genevieve von Petzinger, a graduate student at the University of Victoria in Canada, compiled a detailed listing of the nonfigural images from 153 different sites in France (Figure 22). She was able to group the symbols into 28 different categories, based on their appearance, and to examine their frequency and distribution through time. Such European cave art has always been included as evidence of modern human behavior and symbolism.

Startlingly, the much earlier Blombos cave engravings clearly fit neatly into several of Von Petzinger's categories: lines, cross-hatching, cruciforms, penniforms, and perhaps others. However, the frequency of various symbols at Blombos does not mirror their frequency in Von Petzinger's sample—with the exception of lines, which are nearly ubiquitous—and there are many symbols in the more recent French samples that do not appear in the currently known Blombos sample. Possibly the resemblances are in part due to the simple mechanical issues of engraving hard substances with stones (on ochre slabs or cave walls) versus painting with pigments. However, that the same symbols should recur across such reaches of space and time is astonishing.

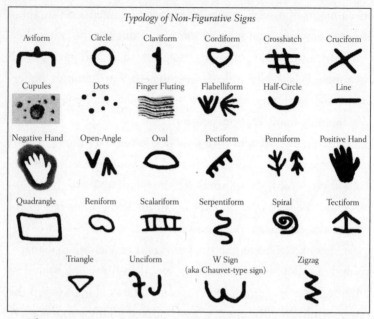

Typology of Non-Figurative Signs

Aviform	Circle	Claviform	Cordiform	Crosshatch	Cruciform
Cupules	Dots	Finger Fluting	Flabelliform	Half-Circle	Line
Negative Hand	Open-Angle	Oval	Pectiform	Penniform	Positive Hand
Quadrangle	Reniform	Scalariform	Serpentiform	Spiral	Tectiform
	Triangle	Unciform	W Sign (aka Chauvet-type sign)	Zigzag	

22. *The geometric paintings and carvings in prehistoric caves from Europe can be categorized into these twenty-eight types, defined by Genevieve von Petzinger. Notice that the carving on the ochre from Blombos Cave would fit neatly into the scalariform and cruciform types.*

What's more, recent discoveries at Diepkloof Rock Shelter in Western Cape Province, South Africa, reveal another time and place where similar geometric symbols have been found: on the rims of ostrich eggshell water containers, some 60,000 years ago (Figure 23). The discovery team, led by Pierre-Jean Texier of the University of Bordeaux, suggests that these engravings constitute a kind of written communication. The same two patterns recur and might indicate something like the ownership of the eggshell or the artist's identity. Finding a way to carry and store water is obviously a vital technological advance for a thirsty animal like a human being. To find evidence suggesting that the containers may have been personal possessions or artworks so long ago is even more astonishing.

Another type of symbolic expression is represented by objects of personal adornment. Recent discoveries have pushed the earliest evidence of such objects back beyond 100,000 years ago. Beads made out of deliberately perforated shells—think of them both as jewelry and indicators of group membership, like gang colors—occur in Algeria and Israel at about 135,000 years ago, and in Morocco and Algeria between 73,400 and 91,500 years ago (Figure 24). The same type of shell beads have been

23. *These fragments of ostrich eggshell from Diepkloof Rock Shelter in South Africa are 60,000 years old. They were engraved with standardized geometric designs that may have indicated ownership. Eggshells were probably used as water containers.*

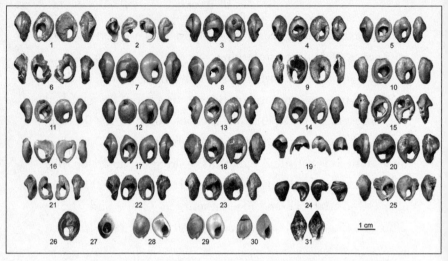

24. The earliest known beads are made from shells of Nassarius gibbosulus that have been perforated with tools. Shells 1–19 are from the Moroccan site of Taforalt (Grotte des Pigeons) at about 82,000 years ago. Similar shell beads are known from less well-dated sites like Rhafas (20–24), Contrebandiers (25), and Ifri n'Ammar (26–27). Shell 28 is a modern Nassarius gibbosulus; shells 29 and 30 are modern Nassarius circumcinctus; shell 31 is a modern Columbella rustica.

excavated from Blombos Cave at about 75,000 years ago. In Mumba Cave, Tanzania (about 52,000 years ago), and Enkapune ya Moto, Kenya (over 40,000 years ago), people were making beads out of ostrich eggshell.

Wearing jewelry—like painting your skin with ochre in a particular pattern—strongly suggests that you are signaling your membership in a particular social group. The same impulse today leads you to wear a uniform that proclaims your job or a T-shirt that sports the name of your favorite band. Personal adornment says: "I know who I am and I want you to know too."

Francesco d'Errico also worked on these beads and finds the invention of personal adornment "one of the most fascinating cultural experiments in human history. The common element among such ornaments is that they transmit meaning to others. They convey an image of you that is not just your biological self." Personal ornamentation or adornment is a form of communication or possibly language.

The need to communicate in this way also implies things about social structure and mobility. If you never run into strangers—into people who have not known you since birth—there is no need to announce your membership in your social group visually. Thus personal adornment hints that humans are traveling more and perhaps more frequently encountering strangers than they once did, which implies an overall higher density of human populations.

Francesco also observes that these beads indicate the probable existence of long-distance trade networks: one of the classic modern behaviors. The shell beads from Morocco and Algeria have been found so far inland that either people made long trips to gather them, or—more probably, in Francesco's opinion—developed elaborate long-distance exchange networks in which coastal and inland peoples traded resources. Evidence of organized networks like this help reveal the connections between ancient peoples, between cognition and culture. The biggest surprise is how early these trade networks apparently existed compared to the traditional expectations of when modern behavior appeared based on European data.

Another criterion is that more modern humans use a wider range of raw materials to make tools, including bone, antler, and ivory. If using bone to make tools makes you modern, then hominids have been modern since 2 million years ago at Olduvai Gorge and at Swartkrans and Drimolen in South Africa—a time when they were distinctly not modern in anatomy. To be fair, the

25. *This finely crafted bone point from Sibudu Cave in South Africa is 61,000 years old and was used as the tip of an arrow.*

"modern" bone tools are far different in shape, function, and general skillfulness of execution compared to the earliest ones (Figure 25). But many archeologists are careless in their language and simply refer to the appearance of bone tools. An expanded repertoire of more sophisticated tool types, such as the production of exquisite and refined bone tools, blades, hafted points, microliths (very small stone tools), and grindstones also occur in Africa at various times, ranging from 280,000 to about 80,000 years ago.

Hunting large and dangerous animals is another supposed sign of modernity that appears on trait lists. Hominids seem to have been confronting large and dangerous animals since stone tools were first invented, judging from the evidence of cutmarks and percussion marks on fossil bones. The issue is not whether animals were scavenged or hunted but that hominids or humans have been dealing with such species for millions of years; it is not a newly acquired or recent behavior from merely 50,000 years ago.

Another trait often put on lists of modern behavior is exploiting a wider range of animal and plant resources, including aquatic organisms and birds. Frequent use of fish and other aquatic resources is not necessarily a late behavior in human evolution, though it has long been assumed to be. Almost 2 million years ago, hominids at Koobi Fora left remains of fish and a tortoise at a site, along with antelope and hippo bones. Similarly, there is considerable evidence that hominids at Trinil, Java, collected and exploited shellfish nearly 2 million years ago. As for exploiting a broad range of animals, remember that animals from Olduvai bearing cutmarks range in size from hedgehogs to elephants. How much broader a range of prey can be possible?

Paleoanthropologists often cite the fact that humans seemed to expand their geographic range when they became modern— but such expansions generally pale in magnitude beside the expansion of the human geographic range from Africa to cover most of the Old World at about 1.8 million years ago. A greater enlargement of geographic range can hardly be conceived.

Controlling fire is another modern behavior—except that Goren-Inbar and her colleagues have shown controlled fire at 790,000 years ago in Israel. The same team has demolished the idea that creating a highly structured living space, with separate areas for separate tasks, is purely modern. They have shown a spatial division of the living area at Gesher Benot Ya'akov which reflects the different tasks performed in each. In short, many of the so-called "modern" characteristics occurred very early in our history and are sometimes associated with nonmodern hominids in Africa. Are these lists of "modern human traits" simply nonsensical? Do the number of attributes that date to deep antiquity mean that hominids were simply "modern" from their first appearance as toolmakers? No, and no.

McBrearty and Brooks are right; using checklists of modern traits may be simple, but it is misleading. Their review shows that all of the criteria of modern behavior are met long before 40,000 years ago, and usually in Africa first. So, why hasn't everyone decided that the previous synthesis concerning the "human revolution" was simply wrong, based on incomplete and Eurocentric data?

The problem is not that the quality of the evidence from Africa is poor or ambiguous; nor do these traits occur as isolated, "one-off" occurrences. The problem is that these signatures do not occur all at once and do not occur all at one place. Art pops up here, disappears, reappears elsewhere much later. Shell beads are 135,000 old, or 91,500 years old, or 73,400 years old; they are Algerian, Moroccan, Israeli, or South African; and they disappear about 70,000 years ago, along with engraved ochre, projectile points, and refined stone tools. The presence of these "modern" traits is neither continuous through time nor clustered in space.

Did people simply make and then forget some of these technological advances? Or did a small group sometimes invent a new tool type or a new form of communication, only to die out without passing on its knowledge?

McBrearty and Brooks suggest that the transition to fully modern behavior was "fitful" rather than occurring as an identifiable revolution. Their conclusion—that there was no human revolution but only a sporadic, here-and-now development of these behaviors—is not universally accepted. Skeptics challenge the dates at one site or the interpretation of specimens at another. They often point out that the first appearances may be early but the coming together of the entire package of traits into a continuous behavior pattern is late, maybe 40,000–50,000 years ago.

McBrearty explains the difficulties succinctly:

> As I see it, these are the challenges to identifying modern human behaviour in the archaeological record:
> 1. the behaviour must involve material objects;
> 2. the material objects must be preserved;
> 3. the objects must be accurately dated;
> 4. the species of the maker of the objects must be correctly identified;
> 5. archaeologists must agree that the objects are the product of behaviour that reveals advanced cognition or symbolic thought.

She adds,

> But perhaps the greatest challenge is provided by item number 5. Henshilwood and Marean have suggested external symbolic storage as a testable criterion for the presence of advanced cognition in the archaeological record. . . . Application of this criterion will exclude from humanity some societies known from the historic and ethnographic records. I sometimes wonder whether the contents of my own tent in Africa, if buried for 200,000 years, would qualify me as truly human!

How else can the African and the European evidence of the origin of modern human behavior be integrated and interpreted?

The simplest, and perhaps most stunning, alternative hypothesis is that the apparent convergence of behaviors 50,000 years ago in Europe exists because that is when anatomically modern human populations got to Europe, each group carrying its different bits and pieces of knowledge and modern behavior. The appearance of a revolution involving the adoption of a modern behavioral package might have been an illusion, caused by an increase in population density in Eurasia that led to a more reliable exchange of ideas and material goods. The apparent revolution might be the swift spreading of ideas from group to group, not the invention of those ideas.

Certainly, modern humans were spreading all over Eurasia by about 45,000 years ago. There are modern human fossils in Peştera cu Oase in Romania (about 41,000 years ago), from Tianyuan Cave in China (about 42,000–39,000 years ago), from Niah Cave in Sarawak (about 45,000–39,000 years ago), to name a few key sites. Then there is Australia, which cannot have been reached without using boats or some kind of watercraft. With their boats, people got to Australia by at least 40,000 years ago, since a skeleton known as Mungo 3 in New South Wales, Australia, is that old. There is other good evidence of human presence (but no skeletons) in Greater Australia about 45,000 years ago from sites like Devil's Lair in western Australia. William Noble and Iain Davidson have argued that the construction of boats and the navigation of open ocean required the use of language. Clearly, if humans got from the Asian mainland to Greater Australia at 45,000 years ago, they must have been spreading across Asia even earlier.

Civilization, modernity, advanced cognitive ability, the dawn of language—whatever you want to call it—may not have arisen in Europe, as Europeans have so enjoyed thinking for the last few centuries. Modern behavior more probably came together as immigrant human populations from Africa met and exchanged knowledge, habits, or ideas in the rest of the Old World.

10

Spreading the Word

THE DEFINITION OF LANGUAGE as a symbolic, communicative medium that functions to transfer information is workable, but there are still some major problems. Now that Sally McBrearty and Alison Brooks have deconstructed the unitary package of modern human behaviors, we have to ask which of all these behaviors marks the rise of symbolism and the appearance of language. And even though many scholars no longer think that modern cognitive and behavioral abilities are a package that originated suddenly and simultaneously in Europe, we still need to ask: What unites these traits? What caused modern human behavior and its premiere product, language, to appear during human evolution?

As I showed in the last chapter, there is a lot of evidence from Africa that symbolic behaviors began to occur in several areas sometime between about 130,000 and 70,000 years ago. Artwork such as repeatedly and deliberately altering ochre or eggshell with geometric markings, and the creation of objects of personal adornment, speak of a certain self-awareness, of a new recognition of "me" and "us" versus "you" and "them." Groups of humans were regularly encountering strangers, apparently for the first time in history. Strangers in this sense are other groups of humans, people who have not known you from birth. Strangers would need emblems that signaled your origin and might wear emblems signaling their own.

A higher rate of encounters with strangers could have occurred because of a lifestyle change in which human groups began to move over larger territories. Alternatively, such encounters might have increased simply because there were more humans on the landscape and population densities were rising. Either of these interpretations would be consistent with the broader range of land and marine resources that were being used by humans when they became behaviorally modern. Communicating through external symbols was clearly occurring, but what were they saying? What do shell bead necklaces or geometric scratches on pieces of ochre mean?

To get at the meaning and perhaps the driving force behind the evolution of language, we need to look at the first symbolic manifestations that we can understand. These are far from being the earliest symbols. They are those amazing artistic depictions of animals—painted on walls, engraved on slabs of stone, sculpted out of clay or rock, carved out of wood. Though there are thousands upon thousands of examples of prehistoric art, the earliest ones are more enigmatic. Only about 35,000 years ago did humans begin creating symbols that we can look at today and understand.

Being in the presence of such wonderful prehistoric art gives me chills. I remember, vividly, going to visit Lascaux, which is about 17,000 years old. It was a sunny, hot summer's day in southwest France. My husband and I were traveling with two Italian friends, visiting sites where important fossils or prehistoric art had been found, and this was the end of our trip. I remember eating well, staying in lovely little hotels, and laughing a great deal. We had seen beautiful prehistoric paintings and sculptures.

Nothing warned me that the last day of our holiday would have a tremendous emotional impact on me.

Since 1963, the cave of Lascaux in the Dordogne has been closed to the general public because the heavy tourist traffic was damaging the famous paintings. Since then, scientists or artists have been able to apply well in advance (six months or more) for permission to enter the cave. At that time, up to five individuals, plus a guide, could enter for thirty-five minutes, five days a week. Now access is even more restricted because humans aggravate the fungus that is slowly destroying the remarkable artwork in Lascaux. We were fortunate to have been granted such a permit and joined a disabled poet for our tour.

The paved sidewalk leading to the cave has broad, shallow steps that slope downward from a grassy area to a large set of bronze doors, which looked as if they belonged on a Hollywood set as the entrance to an Egyptian pharaoh's tomb. Our guide unlocked and opened the massive doors, ushering us from the bright sunlight and warmth into a cool, dimly lit antechamber. Each of us stepped carefully into a shallow trough of formaldehyde intended to kill any algae or pollen that might be clinging to our shoes. The door to the outside was carefully closed before the door to the next stage was opened. The feeling was a little like entering the airlock of an aviary or a butterfly enclosure, taking precautions so that nothing from the outside flies in, or vice versa.

We left the antechamber and walked gingerly down a narrow, sloping ramp—not the original cave floor—lit by low, dim bulbs to prevent stumbling. We held onto a cold, damp, iron railing on our right, which helped the poet navigate the terrain. She was determined to see the artworks she had read so much about, as were we all, but her difficulty in walking made her efforts more apparent. The cave itself was colder still than the anteroom: 52 degrees, as caves always are. The space felt intimate, small but not claustrophobic.

After about 25 yards, our guide stopped us. We stood a moment in total darkness. With a magician's flair, the guide suddenly turned up the lights and we were surrounded by huge, vivid creatures out of the prehistoric past. All of us gasped. We were in the Great Hall of the Bulls.

No images I had ever seen, no study of the brilliant reproductions in so many books, had prepared me for what appeared. The animals were enormous—some nine feet long—and beautiful. There were no petty artistic conventions: no frames shaping neat rectangular spaces, no careful placing of images at eye height, no ground lines, no landscapes, nothing—only vibrant bulls, stags, and horses that swirled and ran and covered the ceiling, the walls, the bumps, the flat places, and leapt out of the darkness at us. I thought for a moment I was going to fall over backward. The paintings were incredibly bold and enveloping (Figure 26).

I barely listened as the guide used a strong flashlight to identify and point out various features and animals. Analysis, interpretation, words were completely unimportant. There was only the darkness, the light, and the vast, compelling animals.

We moved slowly from chamber to chamber, looking and

26. *These paintings from Lascaux in France show a large bull overlapping with several images of horses and (lower right) two stags. Note the fine details: the dapples on the bay horse's rump, the curly hair between the horns of the aurochs. The paintings were made about 17,000 years ago.*

looking, but saying little. The Great Hall of the Bulls merged into the Painted Gallery. A magnificent stag with palmate antlers stood at the front, then we followed a series of horses and cattle down the cave. There was one peculiar beast—known as the Unicorn though it clearly has two horns—that has never been completely identified. It may be a mountain antelope of some kind. The horses were the irresistible Lascaux "Chinese horses," so named because their stout bodies, erect manes, and dainty feet reminded some long-ago prehistorian of ancient Chinese statues. To me—a horsewoman—they looked like ponies or some chunky, small breed such as Icelandic horses or perhaps Haflingers, running one after another down the gallery. The paintings show variations in coat colors and textures, some for winter and others for summer, some with clear dapples. Inevitably, the horses drew my attention as we walked farther and farther along the narrowing gallery. Suddenly, we came upon a turn in the cave and the stunning image of a horse falling down the shaft at the end, showing what would have happened to us had we not been guided. It was a physical surprise, a blow, to see this image, famous and familiar though it was.

We turned and went back, this time taking a turn through the narrow, low-ceiling Lateral Passage to enter the Chamber of Engravings. There were hundreds, maybe thousands of engravings of animals frantically scratched into the cave walls and ceiling, chaotically tumbling one over the other, obscuring, redefining, superimposed with some special meaning I cannot tell. Nearer the ground most of the images are of aurochs— Europe's wild cattle—then above them deer, and above them horses that cover the domed ceiling.

Beyond the confusion of engravings, we descended again— the poet sadly could not follow us here—and came to the narrow, twisting Shaft of the Dead Man. To the left, moving away from the other figures, was a rhinoceros with its tail up, nervously leaving droppings. I have seen exactly this posture so many times in Africa that I could almost hear the plopping of the dung, spread wide by the rhino's swishing tail. The middle figure was a falling or possibly dead man with an erect penis, but he had a strange head with a birdlike beak or mask. There was also a linear object topped with a bird figure that might be a staff or a spear near the man's right hand. Charging the man was a bison, its head lowered to gouge with its sharp horns. An object usually interpreted as a spear sticks into the bison, which seems to have been partially disemboweled, causing its ferocious attack. What does this vignette mean? Is it a moral story, a warning about hunting bison? Or is the true significance of the painting simply its bold being, its mysterious and hugely affecting representation of a dangerous event of great importance?

We turned back through the Chamber of Engravings and entered the wider Main Gallery. There we found more horses, bison, cows, ibex, and a smattering of enigmatic geometrical signs: dots, subdivided rectangles, cross-hatching. These have caused scholars to argue whether they represent nets or traps, mystical symbols, or hallucinations at the beginning of a shamanistic trance. Some signs demarcate the spots in caves with the best echoes and acoustic resonance. There are two massive red and black bisons, pictured tail to tail, and a charming frieze known as The Swimming Reindeer. Only the antlers, heads, and necks of the reindeer are shown, with their chins held up. The place where their bodies would be is occupied by a deep sinuous crack that seems to represent the river the reindeer are crossing.

At the end of the Main Gallery was a long, straight, narrowing passage with few paintings or engravings. It seemed a long way to walk in the dark after the excitement and color of the other areas. Finally, the walls receded and the space enlarged. We were in the Gallery of the Felines. In addition to horses and bison, there were six exquisitely engraved felines that I identified as lionesses or small-maned lions. At the end, like a punctuation mark, were two rows of three red dots, placed one row over the other.

We had viewed it all, but to study each image would take years. We had to leave the cave before our very breath destroyed the art we had come to see.

We left blinking, quietly, a little sadly. For a few minutes, there was nothing to say that could possibly express the experience we had been through. The sheer artistic grandeur of Lascaux was breathtaking. To see it was an incomparable privilege.

On a later trip, I was lucky enough to see the "X-ray depictions" of kangaroos and barramundi in rock shelters in Australia. They are called "X-ray style" because internal organs and bones are depicted as well as the external features. Still, there are dozens of prehistoric artworks on my list of places I want to go. Seeing Lascaux, I understood viscerally for the first time what prehistoric art was about.

Prehistoric art is a communication—a heartfelt scream of meaning—from our ancient past that is so powerful we can still hear it today. I don't know what the geometric signs in Blombos Cave meant, or for that matter the geometric signs in many of the famous later European, African, or Asian caves. However, I know what Lascaux means: Animals. Animals are important, and this is what they are like. You must remember this!

Lascaux and the other figurative art made in prehistoric times is very literal. You look at an eland painting from the Drakensberg in KwaZulu-Natal Province, South Africa, or a sculpted horse from Cap Blanc in France, and you have no doubt what animal is represented. And you see its coloring, its winter or its summer coat, its mating posture, and defensive stance. You see the line of reindeer fording a river, the unmistakable silhouette of a woolly mammoth seen far down the valley, a giraffe's proud neck. You see how the wounded rhino charges, how horses leap from cliffs, how lions put their noses out to roar, where the kangaroo's liver is, and how the snake with its deadly pattern slithers in the sunlight.

Leaving the emotional impact aside—which is hard for me

to do—I can analyze these communications. Taking figurative prehistoric art as a whole means lumping together creations from about 32,000 years ago at Chauvet Cave in France with those from a few thousand years ago in South America. These communications come from every inhabited continent, a great span of time, and an enormous range of peoples. Nonetheless, if I follow my procedure of looking to the oldest evidence for clues to meaning or function, a few vital points become clear.

Prehistoric art is unquestionably communicative in intent. The precise audience for whom the communication was intended is not always clear and probably varied anyway. But the location of many artworks in deep caves and rock shelters suggests that the audience was not general but was a specially chosen segment of the population.

Prehistoric art communicates—transmits meaning—through a shared symbolic vocabulary. In fact, the vocabulary is so fundamental and so deeply embedded in the minds of the human species that we can read a great deal of it today. I find it truly remarkable that I can look at a painting from 26,000 years ago or more and know what the painter was trying to say because I still share that vocabulary.

In prehistoric art, the topic or content being shared is obvious: it is information about animals.

The overwhelming majority of prehistoric art that we can understand in any way depicts animals. Of course there are some sculpted Venus figurines and paintings or sculptures of humans or human-animal chimeras, too. Most of the time, however, animals are what our ancestors were talking about through the external recording devices (paintings, sculptures, etchings, drawings, and carvings) that they devised. Animals and their anatomy and their habits—the fruits of the observations spawned by our ancient ties to animals—are what was so important to record and share that people created these artworks.

To understand how important this information was, think about what is not depicted or rarely depicted. People hardly figure in prehistoric art, not even people that seem to be of special importance—the subjects sometimes called shamans or sorcerers or magicians. So, even though Robin Dunbar and Tim Ingold point out that language today is primarily about social interactions and people, I think the hard evidence of prehistoric art says otherwise. Artists are not drawing interactions or relationships: who was mating with whom, who stole whose best spear, who gave their relatives skimpy portions of the kill, who was a tracker of particular skill, who was singularly fertile. No, the artists were exchanging information about animals.

Looking at prehistoric art also tells us that plants, trees, fruits, tubers, nuts—although they must have been of great importance to the artists—were not the topic of the earliest conversations upon which we can eavesdrop. The number of times plant resources are shown in ancient prehistoric art is very, very small.

What about tools and weapons, those magically transforming inventions of the first stage of human evolution? They are hardly ever depicted. When they are, there is little detail—a straight line to indicate a spear, a curved line attached to a straight line to show a bow. You couldn't possibly interpret such images as instructions for making or using tools and weapons.

Landscapes are very rarely shown. There is a single instance of a stone tablet from Spain dated to 14,000 years ago that is engraved with what the finders interpret as rivers, mountains, boggy areas, and the location of prey species. But other than this tablet, the location of key resources or geographic features such as mountains, rivers, sheltered valley, outcrops of good rock for knapping, or ochre deposits are not recorded in prehistoric art. There can be no doubt that people must have known their physical surroundings, but they did not draw or carve or sculpt them until 20,000 years after they started carving, engraving, and painting the animals around them.

Dwellings or huts or shelters of any kind are absent, though you would think they were important to everyday survival.

And though animals are the favorite general subject of prehistoric art, we see very few insects, birds, fish, reptiles, or small mammals. It is medium- and large-sized animals that dominate prehistoric art. Such animals and information about them must have been supremely important to the artists.

If all of this useful information was too unimportant to be communicated through the medium of art, how much more important the information about animals must have been in comparison!

Consider, too, the cost of prehistoric art. Somebody invested a lot of time, energy, and resources into creating these depictions. To make paintings, pigments were gathered or mined from several different localities and transported over what was sometimes a long distance. The pigments had to be ground into powders—a laborious endeavor—and mixed with the right sort of binders, which probably had to be determined by experimentation.

Clues to how some prehistoric artists manufactured their paints come from a 58,000-year-old factory for processing ochre that was recently excavated in Sibudu Cave in KwaZulu-Natal,

South Africa. Lyn Wadley of the University of Witwatersrand led a team that found 8,000 pieces of ochre at Sibudu in densities as high as 235 pieces per 1 meter square. The occupation level yielded four well-preserved hearths; the upper surfaces of the ash deposits were naturally hardened into clean, white crusts. Hearths can be used to heat-treat ochre, transforming yellow pigment into various shades of red or brown. The hardened hearth crusts at Sibudu were used as work surfaces for grinding ochre into powder or as receptacles for ochre powder. Excavation also revealed used grinding stones, ochre powder, and worked and unworked ochre chunks. Some ochre chunks came from an outcrop a kilometer away, but others were derived from an unknown source. With the addition of animal fat, ochre can be used as a glue; it is also useful in treating hides and is commonly used today in painting decorations on human bodies and ornaments, or on pots, hides, and walls.

Paints also had to be put into leakproof containers of some sort and applied to the rock with specially made tools. While the paintings were being made, oil, fat, lamps, and torches were expended to light cold, dark caves, and sometimes scaffoldings were erected inside the caves. How did the artists live during their creative periods? Presumably someone else supported them because painting a frieze of bulls, sculpting a life-sized bas relief of a row of horses, or engraving the silhouette of a pride of lions into the hard rock took a lot of time. While the artists created, somebody had to be taking care of gathering berries or hunting for game. And somebody may well have had to be minding the babies, too!

The irresistible conclusion, to my mind, is that recording and documenting detailed information about animals was literally a life-and-death matter to these prehistoric humans. What the explosion of figurative art about 40,000 years ago attests to is the fact that the observations which had been accumulating since our intense link with animals was first forged had reached unmanageable proportions. (This "explosion" is of course part of the primary data that led anthropologists to speak of a "human revolution" between 40,000 and 50,000 years ago.) As humans gained more and more information about animals, I believe they became more and more dependent upon the retention and transfer of that information. The value of detailed information about animals went up. That means that the adaptive advantage accruing to those who discovered a method of recording, documenting, and sharing information externally would have been huge. And language was that method.

Does the appearance of figurative art about 40,000 years ago show that language did not arise until about then? No. However, that art convinces me that language had arisen by then, and I wouldn't be surprised if it began long before, perhaps between 150,000 and 75,000 years ago, when humans were making art and jewelry and using ochre. If only we could understand that early art, we'd have a better idea what was being communicated, but I am willing to bet that the message was terribly important.

There is something else, a discovery that has haunted me since I first read a report of it in a scientific journal. Two French scientists, Iégor Reznikoff and Michel Dauvois, investigated the acoustic resonance of painted caves. They traveled slowly through three French caves, singing and whistling through a range of five octaves. They noted on a detailed map exactly where the resonance was best and what notes created the strongest resonant response. When they compared their acoustic map to one showing the placement of paintings, ranging from full-bodied animal depictions to simple rows of red dots, they found an astounding yet wonderfully satisfactory coincidence: most of the paintings lay within 3–4 feet of a spot with good resonance. Places with poor resonance had relatively few paintings.

This finding has huge implications for how cave art was used. The concentration of paintings in resonant areas shows that sound was used along with the paintings. Reznikoff says that "because of the resonance, the whole body is implicated, sometimes in a subtle way. The approach is essentially physical; in this respect, we may say that the sounds and the whole situation are *primitive*. It is indeed a very strong experience to hear in almost complete darkness the cave answer to a sound produced just in front or just under a picture of an animal, a bison or a mammoth." (Italics in original.)

The conclusion seems irrefutable: the images and signs within the caves were placed so that sound—perhaps music— would enhance the experience of viewing them. From my own visits to painted caves in France, I know this same link between sight and sound applies to many more caves than the few Reznikoff and Dauvois tested. A later worker, Steven Waller, studied acoustic resonance at Font-de-Gaume and Lascaux. Waller found that paintings of ungulates (hoofed animals like horses, deer, bulls, and bison) occur more often in areas that echoed sound effectively, whereas the much rarer images of large cats are placed in areas with low reflective resonance. Whatever the occasions were upon which those wonderful paintings were viewed, they would have been unforgettable *son et lumière* experiences.

The finding of flutes dated to more than 35,000 years old in caves near Hohle Fels, in Southwestern Germany, has quieted any skeptics who doubt that humans made and used music during this period, if not earlier. In excavations during 2008 and 2009 at Hohle Fels, a team of researchers directed by Nicholas Conard of the University of Tübingen found four exquisite specimens of flutes, three of ivory and one carved out of the radius (a wing bone) of a Griffon vulture. Most of the flutes are mere fragments, but because the bone flute is almost complete, the others can be reliably identified and reconstructed (Figure 27). The new bone flute is just over 8½ inches long, and five

finger holes show clearly the marks made during its manufacture. This most complete flute—and probably all of them—was played by blowing obliquely into a V-shaped opening at one end.

From the nearby site of Geißenklösterle, another three flutes—including a smaller one with only three holes, made from a swan's radius—have been found. Other flute

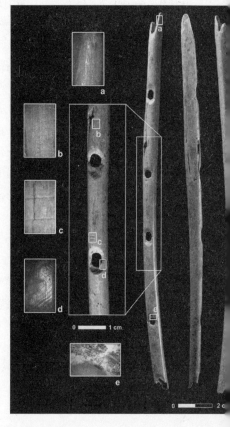

27. *This is one of four bone flutes from Hohle Fels, Vogelherd, and Geißenklösterle, Germany; another four ivory flutes are also known from the same sites. These are the earliest known musical instruments, dating to more than 35,000 years ago.*

fragments turned up at Vogelherd previously, too, bringing the total to eight. The oldest of these flutes comes from layers dated by radiocarbon to about 40,000 calendar years ago, while others range between 35,000 and 30,000 years old. These flutes are currently the oldest unmistakable evidence that humans were making music. Remember that the site at Vogelherd also produced the oldest figurative art in the world in the form of ten beautifully carved figurines representing animals dated to between 31,000 and 36,000 years ago.

These finds suggest that, at the time that cave and portable art was first being made, sound was already being used to enhance the emotional impact of communication. The content that was being transmitted from person to person through the art was also being transmitted through sound, perhaps through songs with words: that is, speech. Thus there is another link between the origins of art and language, and another piece of evidence about the deep significance of communication at this time in human history.

Figurative prehistoric art has the essential components of any communication—audience, shared symbolic vocabulary, and topic—and those of language itself: it is humanly made, communicative, symbolic, social, and (in these instances) written. The two systems seem to have been used to reinforce each other, to embed the information in the mind of the recipients.

Language may have begun much earlier, 100,000 years ago or more, when beads and ochre first began being used. Maybe language stuttered and started, evolved and died out, and then started again in another population for millennia until a critical mass of people with language and a shared vocabulary had been reached.

Theoretically, you don't need a critical mass for language to occur. Theoretically, all you need are two people to create a language. In reality, those people have to be in the same place at the same time and have to have something worth communicating to each other. For language to develop and persist, I think you do need a critical mass of people with enough in common that they have a useful shared vocabulary to use, elaborate, and teach to their children. In other words, until the animal connection enabled human populations to become dense enough on the landscape to regularly encounter one another, lasting language could not evolve. Until that time, language (like making sophisticated bone tools) might have developed and died out only to arise again, leaving a patchwork archeological record of "modern" behaviors.

Let's look at those modern behaviors once again. It is easy to see how using ochre, making art, carrying out rituals, and adorning oneself are linked to language because these are fundamentally communicative actions. What about tool standardization, a broader use of raw materials for tools, hunting successfully and more broadly, controlling fire, organizing one's living spaces, scheduling, making boats, and expanding one's territory? What do these behaviors really have to do with language? The answer is, not a whole lot. These latter traits indicate a certain knowledge base that implies an ability to make observations, draw deductions, and record that information in some means (external or internal) so it can be drawn upon later. Nothing about these latter traits is directly indicative of language and communication. Those behaviors may have appeared or been facilitated by the development of language, but they are not language or communication in themselves.

Looking back at these first two stages of human evolution, we can see the beginning of the animal connection with the invention of stone tools at 2.6 million years ago. At that point, our ancestors came under selective pressure to pay more attention to other animals and gather more information about them.

What we see in the next span of time is the constant racheting up and the growing intensification of the relationship between humans and animals. Knowledge about other animals became more detailed, more voluminous, more important, and more advantageous. Perhaps as long as 100,000 years ago, and certainly by 40,000 years ago, the storing and transmission of that knowledge to others—through language—became a new sort of tool that enhanced human life. Language was not an extrasomatic adaptation but a somatic one, a change that must have happened in our brains and in the wiring among different parts of the brain and among the brain and the lungs, diaphragm, larynx, tongue, and oral cavity.

But language is ephemeral, impossible to hold in your hand, impossible to dig up at an archeological site. Thus the only way we can see language in the archeological and fossil record is to see its extrasomatic manifestations. Words do not fossilize, do not last through time. Sculpture lasts; painting lasts; engraving lasts; and so—astonishingly—does meaning.

11

My Cat Wants You to Open the Door

CLEARLY FULL LANGUAGE, as a means of communicating complex information, would confer a major adaptive advantage upon a social species that relies heavily upon information. Language is a multipurpose tool. It facilitates learning and obviates the need for each individual to observe or figure out every principle, every fact for him- or herself.

Michael Tomasello, an anthropologist at the Max Planck Institute for Evolutionary Anthropology in Leipzig, asks, "What enables human beings to create and use language and other symbols, to create and maintain complex instrumental technologies, and to create and maintain complex social organizations and institutions?" His key point is that individual human beings aren't responsible for these advances. "These are all collective cognitive products," he asserts, "in which human beings have in some way pooled their cognitive resources."

Language is the mechanism that humans use to pool their collective knowledge. Because of language, specialization becomes easier. I can gather berries while you stalk prey, or I can paint pictures of mammoths while you make sure there is something to eat tonight. And, for Tomasello, language is a primary biological adaptation of our species to culture. Language enables humans to transmit ideas as well as facts—ideas such as a better way to make a tool for this purpose, for example—to other members of your group and then from group to group. In this way, innovations and advances persist until further modifications are invented, so the entire culture changes cumulatively, racheting up from the previous situation. It is precisely this sort of racheting up that we see in the fossil and archeological record of interactions between humans and animals.

But humans are not the only social animal that relies on complex information or that would benefit from pooling knowledge. For this reason, researchers have explored the distinctions between human language and communication with or among other animals.

Anyone who owns a pet and pays attention knows that animals can communicate with humans. Indeed, when you acquire a pet, the pet quickly starts training you to understand its needs and desires. You may think you are housebreaking the dog by training it to defecate or urinate outdoors; you may think you are training the cat not to scratch the sofa. In reality, they are also training you to understand their wants and needs—"I have to go out now unless you want me to urinate on this carpet," and, "I'm bored and I'll pull threads out of the sofa if you won't play with me." Pets communicate other needs very effectively, like their wish to be petted or fed or played with, let in and out, taken for a walk, and a dozen other things. Animals with alert and receptive owners can express a wide range of emotions and requests, and the owners indicate their understanding by complying.

Animals in experimental, captive situations are capable of even more sophisticated communication. Educated apes use more than 300 words and rudimentary syntax, but they do not combine more than a few words into sentences. Here again, there are echoes of the finding by Bates and Goodman that—in humans—about 400 words is the crucial threshold of vocabulary size that is necessary before grammatical competency takes off.

Kanzi, the best toolmaking bonobo in the world, is also among the most linguistically adept animals in the world. In fact, it was probably Kanzi's familiarity with learning language from humans—with learning how to learn from humans—that spurred Nick Toth to ask Sue Savage-Rumbaugh if he could try to teach Kanzi to knap stone.

The bonobo language experiment is best understood in the context of Kanzi's interesting personal history. He was born on October 28, 1980, to a captive-reared mother, Lorel. If events had proceeded normally, Kanzi would have returned with Lorel to the San Diego Zoo, but he was born into a bonobo colony where Lorel was on loan for breeding. Because Lorel had no experience in raising infant bonobos, her caretakers were concerned that she might hurt her infant or simply show little interest in him. They watched events closely after Kanzi's birth; Lorel seemed exhausted and confused.

The other bonobos in the colony were keenly interested in Kanzi, especially Matata, an experienced and recent mother who had been raised in the wild. Almost immediately after Kanzi's birth, Matata managed to kidnap Kanzi from his bewildered mother. Even with her own baby and Kanzi to care for, Matata seemed to manage well, and after a few hours, Lorel gave up trying to reclaim him from the higher-ranking Matata.

Matata was already the subject of an experiment in which Sue and her colleagues tried to teach language to chimps and bonobos through the use of symbols. Initially the symbols were printed on paper but the research now relies on computer touchscreens. The symbols were arbitrary and not pictures of objects, though each symbol or lexigram stands for a particular word.

The idea was that Matata would slowly understand that a particular symbol represented a particular object or substance, such as a ball or yoghurt, or an activity, such as tickling or chasing. Sue Savage-Rumbaugh and the other researchers spoke each word aloud and pressed the symbol simultaneously (Figure 28). Because Savage-Rumbaugh had worked previously with two chimpanzees—Sherman and Austin—trying to teach them to communicate with each other through lexigrams, she knew she first had to get Matata to focus on the task at hand.

One of the key prerequisites of any communication is paying close attention to the other individual's actions. As the wife of Willy Lomax demands in Arthur Miller's *Death of a Salesman*, "Attention must be paid." This principle holds true whether communication is between two humans, two apes, or across species boundaries. No one will receive any communication, however earnestly transmitted, unless he or she pays enough attention to know that information is being transmitted.

Savage-Rumbaugh and her collaborators had to be terribly careful to avoid inadvertently cueing Matata about the right answer. The Clever Hans phenomenon has haunted animal language studies for years. Clever Hans was a horse that

28. *Kanzi, Panbanisha, and other bonobos have learned to press lexigrams, such as these shown here, to communicate certain words. The lexigrams are arbitrary symbols. Words used by the bonobos include adjectives, verbs, and nouns.*

became famous in the early 1900s in Germany for his ability to understand human speech. Wilhelm von Osten, a retired schoolteacher in Berlin, trained Hans to tap his hoof a particular number of times to indicate a letter. By tapping in sequence, Hans could perform simple arithmatic calculations, tapping out the answers again with his hoof. He became a sensation all over Europe until a clever test revealed that his trainer was inadvertently signaling the horse when to stop tapping. The Clever Hans phenomenon is often cited by skeptics who don't believe any animal is capable of language.

The tragedy is that Hans was indeed communicating with a human and demonstrating understanding of a rather subtle nature. Was Hans talking? Well, no, not with a voice. But he was receiving a communication from his trainer and responding; he proved that by his behavior. Communication is the basis upon which language is built. To my mind, the Clever Hans phenomenon is not a story about a failure to teach an animal language but rather a success story about achieving a fairly sophisticated level of communication across species boundaries. As my former horse trainer used to say, "Why aren't those people who are interested in the origin of language hanging over the rail, watching people train horses?"

Though Matata was eager to learn, paying attention to Savage-Rumbaugh and the computer keyboard wasn't always easy because six-month-old Kanzi was also in the room, running around, playing, jumping, screaming, occasionally pressing a key, and generally distracting everyone.

By the time Kanzi was two years old, he had learned the meaning of the chase symbol and would invite Savage-Rumbaugh to chase him by pressing the correct key. Matata was having a harder time of it. She learned only six symbols after 30,000 trials and her comprehension was spotty. She could use a lexigram to make requests—pressing the key to ask for an apple—but if the researcher held up an apple, Matata wouldn't consistently press the correct key to name an apple. She seemed to know that pressing keys was desirable but didn't really grasp the object-symbol connection. The Yerkes National Primate Research Center in Atlanta, where Matata and Kanzi lived and Savage-Rumbaugh worked, decided to send Matata back to the field station for a few months to be bred again to Kanzi's father, Bosondjo, while Kanzi stayed behind to begin language training.

The day after Matata left, Kanzi's first training session began and he immediately started using the keyboard. Without deliberate instruction, he had learned what symbols were and knew eight different lexigrams. More important, he understood that the lexigrams were the key to communicating with Savage-Rumbaugh and the other researchers. However, Savage-Rumbaugh found it nearly impossible to get Kanzi to sit still for long enough to conduct careful tests of his understanding or to learn new lexigrams. Perhaps this approach, which is akin to drilling the multiplication tables in math class, and her advanced age (she was ten years old when training started), were why Matata had learned only six symbols in 30,000 training trials over two years.

Savage-Rumbaugh was insightful enough to give up on formal teaching and testing; she decided instead to use games and a language-rich environment to expand Kanzi's linguistic accomplishments and test them. She started taking Kanzi for walks in the woods where seventeen different locations were set up with different foods and games that were played at each. There were also portable versions of the lexigram screens to use. Savage-Rumbaugh created things that Kanzi and she could talk about that would be of mutual interest: what they did or wanted to do each day. This approach was far closer to the way human infants and toddlers learn language; young humans live surrounded by people doing things and talking about events, ideas, objects, and plans, and absorb language spontaneously from their human companions.

Kanzi's vocabulary increased dramatically under Savage-Rumbaugh's new approach. Panbanisha, Kanzi's half sister, and several other bonobos are now being raised and exposed to language following this method. In 2009, when I visited the Great Ape Trust (where they now live) to learn more about this research, there were 384 lexigrams on the touchscreen used by the bonobos.

After thirty years of training, Kanzi's productive vocabulary is still below the 400-word threshold after which grammatical ability soars. Savage-Rumbaugh and Bill Fields, the director of scientific research at the Great Ape Trust, maintain that Kanzi comprehends many more words in spoken English. As among humans, in bonobos comprehension precedes production, according to Savage-Rumbaugh and Duane Rumbaugh. They suggest that comprehending speech is the foundation of language.

Verbal comprehension as an aspect of the bonobos' linguistic competency is currently being measured. In any case, Kanzi has demonstrated that he can understand complex sentences with embedded clauses or recursive structure, something Chomsky asserts that only humans are capable of doing. The proof of Kanzi's understanding lies in his behavior. That he understands communications is shown by his responding correctly to requests such as, "Can you throw the ball in the river?" His response involved not only comprehending the request and acting upon it, but also deciding that violating a long-standing rule against throwing toys into the river was acceptable upon this occasion. Savage-Rumbaugh used behavior as the test of comprehension so that her documentation of Kanzi's linguistic abilities could be more objective.

Kanzi's performance in language production is slightly smaller than a human toddler's—about 400 words, though Savage-Rumbaugh believes he can comprehend about 3,000—and his command of language lags behind what a human toddler can normally produce. At 30 months, toddlers have a vocabulary of about 500 words, and their ability to express themselves, to make complex utterances, and to use grammar is beginning to increase dramatically. In further contrast, the average high school graduate has a vocabulary of about 60,000 words, greatly exceeding ape performance.

The point is not that Kanzi and the other bonobos cannot learn or have not learned language as a human would have. Indeed, they seem to be following the same essential sequence, with comprehension coming first and production following. But the bonobos have not (yet) reached the second crucial threshold of vocabulary size, so it is perhaps not surprising that the complexity of their utterances and grammar lags behind human levels. Possibly utterance complexity and grammar are stuck at a low level and will never progress. The point is that Kanzi *has* learned to communicate with humans at a fairly sophisticated level for a bonobo, showing that at least some of the neural connections and intellectual abilities are there.

Kanzi's sentences tend to be short and, by human standards, awkward. The computer-voiced words in no way resemble a human cadence. Another problem is that using the lexigrams and panels is a cumbersome mechanism at best. Even humans who work repeatedly with the lexigram panels produce robotic and slow sentences often lacking disambiguators. We simply cannot tell if Kanzi would produce more complex sentences if the mechanism were better. Even if all possible disambiguators were present among the lexigrams, I would be tempted, if I were using the device, to leave most disambiguators out in the interests of producing an utterance at something like normal speed. Cross-species communication is simply clumsy. And, of course, humans teaching animals language are attempting to teach them human language, not ape or parrot or dolphin language.

As any animal trainer knows, the initial communication with an animal—- the first trick—is the hardest. After an individual horse, dog, bonobo, or lion learns one specific communication with a human, the second trick is much easier and the third even easier. Vicki Hearne was an extraordinary animal trainer, philosopher, essayist, and poet. Before her death in 2001, Hearne understood the process of communicating with other species as well or better than anyone else. She refers to this "learning to learn" period as establishing a set of common values, in which the animal comes to believe that trying to communicate with humans is worthwhile and meaningful.

Like many animal trainers, Hearne referred to the communication between animal and human using the metaphor of speech. Dogs, horses, wolves, cats, and other creatures are often said to "talk" to their humans, though verbal communication is not all that is meant. This sort of "talking" may as often involve body language or gesture as an utterance. Calling this cross-species communication "talking" is the shorthand of our lazy, speech-dominated species.

One of Hearne's key points is that this remarkable ability for cross-species communication is built, first, on a recognition that communication is possible. If an animal—or human—does not think that communication with the other is possible, or interesting, no attention will be paid and no messages will be sent or received.

Here is one of the unexpected and extraordinary differences between humans and chimpanzees. Chimpanzees are rarely observed to actively teach another anything: only a few instances have been noted in decades of chimp-watching. Chimps—and bonobos—do demonstrate or show things to others, but they do not engage in the encouragement that humans regularly use in teaching. Of course, much of human encouragement is verbal, which in nonverbal animals can't possibly occur. But there is also modeling, in the sense of helping another use its hands (or legs or feet or whatever) correctly by physically correcting their position. This too seems rare among chimpanzees.

Although infant chimps watch their mothers intently, the mothers do not—remarkably—watch their infants intensely or try to help them learn important skills. After studying how young chimpanzees learn to crack nuts using stone anvils and hammerstones, Noriko Inoue-Nakamura of Kwansei Gakuin University and Tetsuro Matsuzawa of Kyoto University commented that, in their study, the mothers showed no active teaching of their infants nor any social reinforcement for correct behaviors. Shockingly (to a human), "the chimpanzee mothers' attitude toward their infants was characterized by the lack of any feedback to infant chimpanzees' attempts at nut-cracking behaviors."

Is this lack of attentiveness to others an underlying factor in chimps' lack of language? Or is teaching simply performed at a more subtle level? Sue Savage-Rumbaugh claims bonobos teach each other by subtle demonstration and response, not by didactic instruction, and the same might be true of chimps. Yet Barbara J. King of the College of William and Mary has analyzed ape communication as a co-regulated dance, a complex creation of meaning through body language and gesture. Her work emphasizes the importance of each animal's awareness of the other, in subtle reciprocity. And Jill Pruetz has seen mother chimps in Gabon showing their offspring how to make a spear, how to stick it into a tree hole, and how to smell the end of it afterward, to see if the spear has hit anything.

Certainly, chimps and bonobos can learn many of the rudiments of language with training and human intervention. And, as Savage-Rumbaugh and Hearne both know, the first step toward training a bonobo (or any other creature) is to capture its attention. Without attention, mutual attention, communication is simply not possible. Neither, Tomasello would argue, is the creation of culture with its characteristic racheting up.

The second crucial point is that humans often fail to appreciate the significance of communicating with another species. This is perhaps why some linguists think the Clever Hans phenomenon is a "mere trick," not a tremendous breakthrough. It is true that trivial remarks addressed to an animal—"Who's a pretty doggie then?"—may not be understood precisely and the lack of communication matters little. But sometimes something truly important is said and understood, something perhaps that says to the other being, "This is the kind of animal I am, this is what is essential to me," or, "If you do this, I will probably die"— a thought that frequently occurs to novice horseback riders. When communication is about something vital and urgent, it is of enormous advantage. But the ability to communicate with another species is perhaps an even bigger and more important accomplishment, for it enlarges both species intellectually and emotionally.

To arrive at meaningful communication, both human and animal have to send out signals on a vague hope, like beaming radio signals into deep space over and over on the chance that aliens will decipher them and answer us. Of equal importance is recognizing the responses when they come, if they come. A trainer may utter the command, "Truro, *sit*," to the new puppy while physically urging the dog into a sitting position many times before Truro grasps the concept that *sit* means the physical action. Once that possible connection crosses Truro's mind, he will show his understanding by voluntarily sitting in response to the command

and then looking for a response, for approval. "Truro, *sit*," may be uttered many more times, in addition to the reward of "*Good dog*" or a pat, before the concept of what a formal *sit* entails is grasped. In dog training, a formal *sit* is immediate, precise, symmetrical. It is as different from simply sitting down as standing at attention is from simply standing. A formal *sit* in dog training expresses readiness, attention, willing obedience, and the expectation of a job to be done. The first formal *sit* a dog correctly performs intrinsically asks, "Is this what *sit* means?" The only possible answer, if training and communication are to continue, is, "Yes, that is the essence of *sit!*" If the moment is missed, or not granted its well-earned significance, training will take much much longer and may never succeed.

In that glorious instant when a human and an animal converse respectfully, appropriately, and in full awareness that a dialogue is taking place, something magical happens. Of course, sometimes the animal initiates the conversation and it is the human who replies. If you have ever been involved in communicating with an animal, you know what I mean. No matter who opens the dialogue, that instant of genuine communication shows that you and a member of another species have found a shared value, perhaps the common value that shapes each of their very beings. You have found a way, however fleeting, to say something meaningful to a member of another species and to hear what that animal replies. Such interactions are astonishingly compelling and powerful.

What actually happens when Kanzi presses a series of lexigrams to express his interest in going to a particular place to look for M&Ms? What happens when he hands Savage-Rumbaugh the red hat, not the blue hat, as he has been asked to do?

What happens is communication: a real, unmistakable transmission and reception of meaning between bonobo and human. That this communication can take place—does take place—between two species is remarkable. Yes, to reach this point took months of training and intensive interactions of the sort that are most unlikely to occur in the wild. Nobody claims bonobos have language in the wild or that they use language to communicate with humans in the wild, either. Nonetheless, that moment when Savage-Rumbaugh and Kanzi first knew what was going on, first realized that they were communicating, must have been extraordinary.

A very similar sort of breakthrough is dramatized in William Gibson's play, *The Miracle Worker*, which tells the story of Helen Keller, a girl who was deafened and blinded by scarlet fever at the age of nineteen months in 1882. Unable to communicate with her family or to understand them, Keller grew into a wild and uncontrollable child. When she was six, her parents persuaded Annie Sullivan to work with Helen. Annie was a former pupil of the Perkins Institute for the Blind; she was visually impaired but adept at speech and sign language. Annie began spelling words into Helen's hand and trying to get the child to make an association between these novel sensations and the objects they described. The climax of the play occurs when Helen suddenly makes the connection. In her autobiography, Helen wrote:

> I stood still, my whole attention fixed upon the motions of her fingers. Suddenly I felt a misty consciousness as of something forgotten—a thrill of returning thought; and somehow the mystery of language was revealed to me. I

knew then that "w-a-t-e-r" meant the wonderful cool something that was flowing over my hand. That living word
awakened my soul, gave it light, hope, joy, set it free! There
were barriers still, it is true, but barriers that could in time
be swept away. . . . Everything had a name, and each name
gave rise to a new thought.

Though the breakthrough in part may have been due to Hel-
en's ability to speak prior to her illness, the recounting is deliberately dramatic. But there can be no doubt that the origin of
language contained many such moments when two individuals
found themselves suddenly able to share information. Naming,
for Helen, was the first step in having power over her world and
being able to control herself and her life, as it is for many hearing children.

A closely parallel realization occurred to Genie, the abused
child who was isolated from human communication until age
thirteen. As recorded in books, articles, and videotapes, as soon
as Genie realized that there were names for things in the world
and that she was permitted to voice them without punishment,
she became hungry to learn the names of as many things as she
could. Her joy in learning the words and experiences of which she
had been deprived for so long seems like an intellectual explosion.

Neither Genie's experiences nor Helen's tell us how humans
first invented or evolved language. What they show us is how
immensely powerful the ability to encode and transmit information is to modern humans.

In our evolutionary history, it seems inevitable that the ability to communicate more and more effectively spread rapidly
after such breakthroughs. And this new and powerful ability—

language—altered the life of those who possessed it. As Chomsky argues, language is used in part for ordering and developing thoughts.

Language also has important social and economic functions for working and living in groups. True, language is not necessary in order to work cooperatively with another individual. Lions lack language (as far as we know) and yet hunt cooperatively all the time; many species share tasks like babysitting or standing guard while the rest of the group forages without any explicit linguistic agreements that humans can detect. Many primates also lead complex social lives and keep close track of shifting alliances and the ever-changing positions of the members of a social group, without language. Though communication is clearly possible and extensive without language, planning, bonding a group into an effective social unit, and complex cooperation are surely facilitated by language.

Creating a division of labor and developing specializations would also be enhanced by language. For example, a better hunter might trade meat in exchange for better stone tools that were skillfully knapped by another; artists recording valuable information about the habits of game animals in paintings and carvings might be able to add information based on others' knowledge or experience. Language meant it was no longer necessary to rely only on what you yourself saw or heard; my information could be supplemented with your experiences, opening up a revolutionary new panorama of opportunities.

Language triggered the birth of a new world—for humans and for the animals around them—because language allowed humans to talk about animals and in time, with animals.

12

Living Together

LEARNING HOW TO COMMUNICATE with other humans in the complex and sophisticated manner we call language was a stunning accomplishment. Some of the selective advantages that accrue to those with language are based in part on simply learning to pay close attention to other individuals and in part on what psychologists call the Theory of Mind (ToM).

At its essence, Theory of Mind is the awareness and appreciation that others have minds, emotions, and sets of perceptions that are (or may be) different from our own. ToM is the bedrock upon which any attempt to explain, predict, or influence the behavior of others is based. ToM is what makes communicating with another human worthwhile—even evolutionarily advantageous—because that other has experiences, knowledge, or ideas different from your own. And, as the discussion of the origin of symbolism and language in the last chapters showed, much of the most valuable information exchanged among our ancestors at the beginning seems to have been focused on animals.

The last great stage of human evolution began about 32,000

years ago. This final stage is marked by a vastly different perception of others, both animal and human. And this perceptual shift caused dramatic changes in the world.

The late, great archeologist V. Gordon Childe proposed in 1936 that this last of the big behavioral leaps in human history—he coined the phrase the "Neolithic Revolution" to express his idea—came about when humans in the Near East began domesticating plants and animals and by so doing invented agriculture and a new lifestyle. He believed this revolution occurred starting about 10,000 or 12,000 years ago and spread outward from the Near East to the rest of the world. As a graduate student in anthropology, I learned about Childe's Neolithic Revolution and marveled at the way this one synthesis explained so much.

Childe observed the profound effects of domestication. Having a staple crop like corn (maize), rice, barley, or wheat that could be raised in large fields meant that humans could live in high densities—settlements like towns or even cities—and thrive. In fact, having domesticated crops dictates that humans need to live in more fixed and larger settlements. Simply planting seeds and going away to return in a few months in hopes of finding a crop to harvest doesn't work. Plants need to be tended, watered, perhaps fertilized, weeded, protected from animals that might devastate an entire field in an afternoon, and defended from other humans who might do the same. Once humans have crops, they need to settle down at least for the duration of a growing season. As Childe reasoned, building more permanent shelters instead of transient camps made sense for farmers, and the archeological record as it was known in his day verified this logic. Settling down could be done singly—a field here tended by one family, another a few miles away tended by another, and so on—but farming or gardening works better when human groups combine and work cooperatively. This fact means that most agricultural or even gardening peoples live in villages or towns, not as scattered individuals.

According to Childe's view, plants were domesticated first and tied people to their fields. Crop farming works neatly with keeping small stock animals, like sheep and goats, which can be fed on the rubble left behind after a field has been harvested. Mixed farming was a viable and effective means of gaining new control over food resources and this change is at the heart of Childe's vision of Neolithic Revolution. People began living in larger groups to support this new economy. In turn, the ownership of land—or at least having rights to whatever was produced on it—became crucial. Herds could be moved from area to area, but crops could not be moved until harvesttime. Who would work the land, fence it, and carry water to it unless he was guaranteed the food he grew on it or the animals he fed on it? Living in large groups with land rights required rules of behavior and rulers—and eventually taxes to support the priests or kings or chiefs who set down the laws and enforced them.

I want to diverge from Childe's theory briefly to clarify what domestication is and isn't.

In animals, domestication is not taming. A wild animal may be tamed during the span of its lifetime and may make a satisfactory pet, though severe problems often arise as wild pets reach sexual maturity. Taming can be thought of simply as behavioral training: housebreaking (if the animal lives in the house), teaching an animal not to bite, training it to accept human presence and touch, getting a wild animal used to eating food provided by humans, and so on. The difference between taming and domestication lies in the permanence of the change in the wild animal. A tamed zebra, or hippo, or coatimundi will not give birth to offspring that are already tame and receptive to human handling. Tamed animals learn particular behaviors but do not pass this learning to their offspring.

In contrast, a domesticated animal species has actually undergone a change in its genetic makeup over generations because of human actions. "Domestic"—like "domicile"—comes from the Latin *domus* meaning "home" or "house." The very term implies that a domesticated animal lives with humans, in their home, not for a single generation but forever.

Though we have some direct evidence of this process of domestication—such as sites where apparently wild animals lived in close proximity to humans—we can only speculate about the thoughts of the humans involved. Perhaps they knew what they were doing, but more probably they were unaware that their choices would have such long-reaching effects. They simply operated by instinct, getting rid of the individual animals that were difficult or dangerous and keeping the ones that were more useful and pleasant. In the case of other eventually domesticated species—like goats, sheep, cattle, horses, cats, and llamas—humans selected against those that were sickly, didn't yield good milk or wool, couldn't stand confinement, were aggressive and troublesome, or simply weren't cute enough. They may also have decided arbitrarily that particular proportions or a particular coat color was preferable, for example. The animals that humans didn't like were killed, eaten, or not allowed to breed. The ones with desirable traits were allowed to breed, and sometimes their sexual partners were chosen deliberately by human owners with the idea of perpetuating a desirable trait.

What about plants? Plants hardly need taming. People probably started tending plants and eventually domesticating them in a rather casual fashion, perhaps planting some leftover seeds near the dwelling for convenience and then creating a more recognizable garden. A little more water and attention helped plants to survive and yield useful food. In time, greater familiarity with the plants might lead humans to select the seeds of those that survived better or yielded better crops in preference to other seeds, if any seeds were left over at all. If everything is eaten, there will be no seeds for the next planting.

And, as David Webster, my colleague at Penn State, points out, there is a paradox involved in domesticating plants. In most cases, a plant was domesticated because its seed or root was edible and could be stored. However, if the reproductive organ of the plant is the same as the food product, saving seed for a future planting compromises the effective yield in the current year. For example, for maize farmers to break even and continue to plant their fields, they must save about 30 percent of each crop for the future. The kernels are the source of food for the human farmers and, simultaneously, the source of future plants. Certainly in the early stages of domestication, the best or most desirable plants were those that yielded the most food. However, the individual plants that are most of the world's staples (potatoes, rice, yams, wheat, maize, barley, or other cereals) reproduce only once. The exceptions to this rule are trees that produce fruits or nuts, which take much longer to reach reproductive age than annual plants, but which also reproduce year after year.

In fact, most plant domesticates cannot survive without human intervention. They no longer self-seed because humans selected for plants in which the seeds remained closely attached to the stalk for easier harvesting.

David realized that, ironically, the offspring (the seeds or nuts) of the best plants are the very ones that were likely to be eaten preferentially by the farmers instead of being saved for the future. After all, who would eat the small, sparse, wrinkled, or discolored kernels of one plant when another in the same field yielded big, fat, tasty-looking kernels? This type of human choice meant that the less attractive and less desirable kernels, which perhaps came from plants with a lesser chance of survival or lesser productivity, were the ones that were left over for future plantings. By picking the "best" kernels—to eat—humans may have delayed the domestication of maize and other grains because eaten kernels (seeds) cannot reproduce.

To some extent, the same issue arises during the domestication of animals if those animals were domesticated for meat. The young animals who put on weight (meat and fat) the fastest might well be eaten first and preferentially. Their individual genes might never be passed on to future generations.

However, the telling difference between most food crops and most meat animals is that the animals can reproduce several times, often annually during their lifetime, and still be edible at the end of their productive years. You might eat the best of the young stock, but their parents could live on to reproduce again. To maintain a roughly constant herd size, a stock farmer must slaughter about 30 percent of his yearlings for meat and keep 70 percent of his young stock for breeding. What a contrast with the grain farmer, who eats 70 percent of the harvested cereals and saves only 30 percent of the seed for future plantings!

The different biology of plants and animals dictates these differences in farming practices. The seeds that are kept for future planting all have a potential to yield food (more seeds) in the next generation, though some carry genes that will produce a more abundant or desirable crop. In contrast, the animals that are kept for future reproduction are specifically selected by sex. Stock farmers usually keep a small number of intact males for stud purposes and also many more females of various ages to produce a good group of offspring the next year. If the animals have additional uses other than yielding meat—such as pulling plows or hauling loads—then castrated males may also be kept. Because many stock animals remain fertile for several years, their young are in some sense a renewable resource.

The process of domestication may have been radically and importantly different if the target species were an animal, not a plant. Humans rarely get emotionally involved with plants, probably because we don't perceive responses from plants to be emotions or individual recognition. Plants don't wag their tails when they see you and plants don't bring you a favorite toy and drop it at your feet in hopes that you will play. Plants don't run over to their caretaker in hopes of getting more food or scratches. In fact, plants do not recognize or love their caretakers, or if they do, they express their affections in ways that most of us cannot perceive.

Humans are more emotionally involved with animals because they and we communicate emotions or opinions through the same mechanisms: body language, voice, and sometimes scent (pheromones). This comes from our shared evolutionary heritage as mammals. Although we know that plants do communicate with one another chemically, we humans are very poor at detecting those communications, much less deciphering them, because we do not share the mechanism. Human simply do not have an efficacious system of communicating with plants, and we do not receive or recognize plant-to-human communications if there are any. You can talk to plants, and many people do, but the plants don't talk back.

Because we have the potential to communicate much more directly with animals than with plants, the process of domesticating an animal is much more intimate, personal, and psychologically powerful than the process of domesticating a plant. In fact, I suggest that the process of domesticating an animal is basically the process of creating a genetically encoded potential for a mutual language or communication system, based on a set of shared values. The possibility of such interactions makes the domestication of animals a quite different task and experience from those involved in the domestication of plants.

A third key difference is that domesticated animals often literally live in the dwelling in physical contact with the family, whereas plants live outside and often in fields some distance away. This spatial separation tends to make the human-plant relationship more distant than that between animals and their owners. The physical closeness—the hugs, the pats, the curling up together to keep warm—are extremely important modes of communication between animals and humans. Plants offer no such communication and don't, as far as I know, enjoy being hugged. Although the raising of domesticated plants plays a more dominant role in providing food for humans than raising domestic animals, we are physically and psychologically much closer to domestic animals.

For all these reasons, and probably more, I do not believe that domestication is a single process with a single effect on

either the domesticator or the domesticated. Though we unfortunately have only one word—"domestication"—to express the
two processes of domesticating plants and animals, they are
vastly different in technique, emotional impact, and outcome.

Not all humans are equally adept at understanding or
handling animals, even if they understand that, for example,
protodogs or early domestic dogs offer useful traits. Once the
domestication process was underway, anyone better with animals in various ways—for example, more skilled in managing
animals, in communicating with them, in assessing what animals needed in terms of care or when they were sick—would
have a natural advantage over the other humans.

Returning to Childe's hypothesis about the Neolithic Revolution, we can see the plausibility of his ideas; but was he correct? To be fair, he formed his hypothesis decades ago and there
have been all kinds of discoveries since. Few grand syntheses
withstand nearly eighty years of new techniques and new evidence. And, whether the Neolithic marks the beginning of
human modernity or not, was domestication really the beginning of the close connection between animals and humans?

As we have seen in the previous chapters, humans were intimately involved with animals and their habits and behaviors for
millions of years. The animal connection has a deep antiquity in
human evolution, and the selective advantages accruing to that
connection explain at least the first two big developments in our
evolutionary history: the origin of tools and the origin of language.

Could something like domestication, which occurred only
yesterday in evolutionary terms, also be part of the long trajectory I have been tracing? I think so. Animal domestication is
another means of making tools, living tools, that perform work

for humans—whether it is by hauling, barking, transporting, or making food out of inedible substances.

Therefore, I question one of the fundamental premises of Childe's Neolithic Revolution: that learning how to domesticate other species, whether plant or animal, was a single spark that set off a revolution. Childe's grand synthesis is an attractive and well-told story, but it doesn't meet the facts as we know them today. It is more a Just So story than an account of the evolution of human behavior.

For one thing, Childe was certainly mistaken about the source of domestication. He believed the entire process started in the Near East, from which the ideas and products diffused outward across the rest of the Old World. Recent fossil and archeological remains of plants and animals show compellingly that different species were not all domesticated in the Near East. Domestication happened not once in one place but many times in many places. For example, yams were domesticated in New Guinea, sunflowers, moschata squash, and maize in the Americas, several kinds of wheat in Turkey, horses in Kazakhstan, and so on. There was simply no single, great center of domestication in the Fertile Crescent, as Childe thought, but many different centers on all continents except Antarctica. Apparently, different groups developed the concept or concepts of domestication and created their domesticates separately and in parallel. Domestication did not spread by diffusion of ideas or techniques through direct contact but arose independently in different parts of the world.

Childe also got the sequence wrong, according to more recent discoveries. He thought that inevitably plants were domesticated first and animals followed later, allowing the domesticators to drive stock animals into harvested fields to eat the remaining stubble. After the animals cleared the fields of refuse, they could be herded to other protected pastures, leaving the fields newly fertilized with dung and available for planting. Economically, this practice makes sense and works. Historically, however, the sequence is wrong because we now know that the domestication of animals came first, well before the domestication of plants.

Was the Neolithic, in fact, a revolution? Childe's trademark term implies that domestication of plants and animals and the subsequent economic and social change occurred rapidly and inevitably. Paradoxically, Childe understood that domestication was a process, not an event, yet he characterized the domestication of plants and animals as a revolution. He wrote:

> The first revolution that transformed human economy gave man control over his own food supply. Man began to plant, cultivate, and improve by selection edible grasses, roots, and trees. And he succeeded in taming and firmly attaching to his person certain species of animals in return for the fodder he was able to offer, the protection he could afford, and the forethought he could exercise.
>
> The two steps are closely related.

We know from hard evidence that the process of domesticating species was not rapid, either as an entire process or on a species-by-species level. Domestication must have been one of the slowest revolutions on record, since early domesticates of various species show up over a span of tens of thousands of years.

Just how long does domestication of an animal species take? No one knows. A few experiments have been carried out to answer this question, most notably one in Russia on domesticating silver foxes, a color variant of the red fox, *Vulpes vulpes*. Starting in 1959, a Soviet scientist, Dmitri Belyaev, established a colony of captive silver foxes in Novosibirsk, Siberia. The animals were kept in cages and the tamest were selected for breeding each year on the basis of their behavior. Those chosen for breeding were the most friendly toward humans, the least aggressive toward humans, and the least fearful of novelty. The animals were handled very little to prevent deliberate attempts by handlers at taming their individual charges. Only 4–5 percent of the male foxes were allowed to breed each year and 20 percent of the females. This regime was the equivalent of a very harsh selective pressure in the wild.

After ten generations, 18 percent of the foxes sought human contact and showed little fear. After thirty-five generations, forty years, and some 45,000 foxes, a "domesticated fox" was created. The domestic foxes were not only tame and people-oriented from puppyhood but, because of the intense selection to which they had been subjected, they were also genetically and physically distinctive. They looked more like dogs, with floppy ears, piebald coloring, and curly tails; they had narrower skulls and shorter, wider snouts. Their neurochemistry was also distinctive. The domesticated foxes' adrenal glands made less corticosteroids—the "fight-or-flight" hormones—producing a basal hormone level that was only about 25 percent of that observed in wild foxes. A related change was a delay in the period of development when they began to show fear of novelties. The domestic foxes became fearful three or more weeks later than in wild foxes, giving a longer window of development during which they might bond to humans. These unexpected traits were not selected for but simply appeared in the population, either at random or because they were in some way linked to the behavioral traits that were selected for.

Does this experiment mean that you or I could domesticate a new species within our lifetime? It's not likely because the experimental conditions here were very artificial. For starters, the original 130 foxes were taken from a fur-breeding farm and so were tamer and less aggressive than wild foxes. Since all of the captive foxes were fed and cared for, there was no natural selection related to such traits as hunting ability, strength, or aggressiveness toward other foxes. Circumstances that might arise under natural conditions and favor more aggressive or more wary animals were eliminated. Finally, the breeding of the foxes was closely controlled, so there were no breedings back to the wild type, which might reintroduce aggressiveness. In fact, the least tame foxes weren't allowed to breed at all.

Under more natural conditions, with animals barely controlled by humans who were not intentionally trying to domesticate them, few if any changes in a wild species could have occurred in so short a time span as forty years. By the time the effects of domestication of any species are visible in the fossil or archeological record, the process has probably been going on for thousands of years.

This observation leads us to another crucial point: domestication began and was recognizable much, much earlier than Childe thought. Until 2009, most paleoanthropologists would have cited 12,000 or 15,000 years ago as the earliest date at which there was good evidence of any type of domestication. Rye is currently the oldest known plant domesticate, first appearing at about 13,000 years ago in Turkey, with emmer and einkhorn wheat following at about 11,000 years ago. Such cereals provide dietary staples that can be sown, harvested, eaten, and stored effectively. They clearly are integrally related to the origin of permanent or semipermanent villages because only such a staple food source can support such settlements. The domestication of animals such as sheep and goats makes the domestication of grain plants even more efficient because even the inedible (to humans) parts of the plants are used. But goats were not domesticated until 10,000 years ago, 3,000 years later than the earliest grains.

Bruce Smith of the National Museum of Natural History at the Smithsonian Institution has spent his career studying the origin of agriculture. He realizes the flaws in the overly simple view Childe had of the process. "The transition from hunters and gatherers to agriculturalists is not a brief sort of thing," he remarks. "It's a long developmental process." In the Near East, people continued to hunt wild game long after they developed domesticated crops—and they didn't always live in permanent settlements, towns, or cities.

But the biggest problem with Childe's chronology and theory becomes glaring when you try to apply it to animals. Childe proposed that the heart, the very essence of the Neolithic Revolution, was that humans obtained greater control over their food sources.

The problem with this idea is that the first species to be domesticated was not a plant that provides a staple crop nor was it an animal that is practical as a food item. The first known domesticate was the dog, and its fossil remains are 32,000 years old, long before the domestication of crops or other animals.

13

The Wolf at the Door

If the heart of the Neolithic Revolution was finding ways to ensure a steady food source, as Childe maintained, then the first domesticated animal should have been one useful as food. It wasn't. The first domesticated species was a dog.

I don't mean by this that humans cannot eat dogs; of course they can. I don't mean that no human culture does eat dogs; they do and have, both in the present and in the past. What I mean is that the point of domesticating dogs was not to obtain a reliable supply of meat, even if early protodogs or dogs were sometimes eaten.

On what basis can I make such a bold statement? On my knowledge of canids, the zoological group to which dogs, wolves, coyotes, jackals, and other similar mammals belong. Canids flatly lack the attributes of good food species and for this reason they are not common food items.

For example, canids don't live happily in flocks or groups of several dozen or hundred individuals, as wild sheep or bison do. Wolves tend to live in packs of from five to about fifteen individuals, depending on the number of pups born in a given year. Similarly, the Cape hunting dog, *Lycaon pictus*, typically lives in packs of six to thirty adults, with a variable number of pups. As a rule, only the dominant female canid in a natural pack breeds in any particular year. In contrast, among herbivores like sheep, goats, or cattle, it is a common ecological strategy for many females to give birth within a restricted period, usually at time during which food is most abundant. I don't mean to imply that the wildebeest or buffalo or goats sit down and strategize or plan their birthing times, but the ecological effect is that the predators taking the newborns are swamped because there are more newborns than can possibly be eaten. Thus, this simultaneous birthing scheme means that enough herbivore newborns survive.

Because canids need large home ranges in which to hunt for prey, they are usually highly territorial and defend their territory fiercely from other packs of canids. If prey are relatively rare and are concentrated in predictable areas during the winter, wolves may relax their territoriality somewhat, but tolerance of alien wolves is of limited duration (weeks or months), at least among living wolves. Since animals raised for food are generally confined and raised in the largest numbers possible in the smallest feasible area, canids do not thrive under this style of maintenance.

Also, dogs are expensive as stock animals, because they eat meat rather than plant foods which occur in large supply. Feeding an animal meat to grow more meat is a foolish idea. Through digestion and other life activities, 90 percent of the energy in the meat that you feed your potential domesticate is wasted as it is turned into new meat.

The differences between raising herbivores (plant-eaters) and carnivores (meat-eaters) become really striking when you compare the economics of dealing with a wild wolf and a domestic goat raised by traditional pastoralists in Africa. (It would be unfair to compare keeping a wolf with keeping a goat under the carefully managed conditions of many farms in the Western world that use specially formulated feeds, modern veterinary care, and other tactics to increase yield.)

At maturity, a gray wolf female weighs 50–85 lbs and a male 70–110 lbs, although wolves living farther north are slightly bigger. A mature goat in sub-Saharan Africa weighs about 80 lbs, so the weights of the animals are roughly comparable.

A goat eats about 30 lbs (liveweight) of combined grass or browse per day: a quantity almost equal to half of the average adult weight of such goats. A wolf needs only an average of 5 pounds of meat a day to maintain sufficient health for breeding: a much smaller percentage of its body weight. However, what the goat eats is generally considered inedible by humans and it is also free, assuming there is sufficient pasture land. The meat the wolf or protodog eats has to be hunted, so it takes human effort to obtain this food. Besides, if you had 5 spare pounds of meat a day, why would you feed it to a protodog?

Another problem is that carnivores as a rule don't mature to reproductive size or full body weight quickly, a desirable trait in stock animals. How do these two species stack up in terms of growth and development?

Pregnancy lasts 56–70 days in goats, roughly the same as in wolves. Twins are common in goats and many females in a herd give birth at the same time. Wolves have litters of four to six pups, but normally only one female per pack breeds in any particular year. A newborn goat gains almost 9 ounces a day for the first twelve weeks of life, more than twice as much as a wolf pup, which gains only 4 ounces a day. A goat is weaned and eats independently when it is only 5 months old, versus the slow-maturing wolf that cannot join the pack in hunting forays until it is at least 7–8 months old. The young goat is sexually mature by 10 months or earlier and usually has its first birth at 14 months. A wolf reaches sexual maturity at about 22 months—taking more than twice as many months to reach maturity as a goat—and often doesn't reproduce at all until it is between 2 and 3 years old.

Thus, if your goal is to care for animals in order to have meat, goats are a far more sensible choice than wolves. Goats grow faster on food that is useless to humans and reproduce sooner. In addition to being slow-growing and needing expensive food, wolves are potentially dangerous to humans and are difficult to keep in large numbers.

No, wolves are quite simply not good meat animals. No one with a rudimentary knowledge of canid behavior and ecology would choose to raise them for meat. No wonder we find absolutely no archeological evidence that dogs or other canids were ever kept in large numbers and killed for meat. Docility, rapid maturation, packing on muscle mass while young, and living in very large groups are the sort of features that you see in animals destined for the dinner table, not ferocity, slow growth, small group size, and territoriality. Because they were canids, early domestic dogs were uncommon and usually found singly. The scientific name for the first species to have been domesticated, the dog, is *Canis familiaris*—the familiar or friendly dog. But they didn't start out as dogs; they started out as ferocious, predatory wolves.

So, why dogs?

How do you transform a wild species like a wolf into a reliable domesticated pet? We can imagine a plausible scenario, for which we have precious little direct evidence. Perhaps human hunters noticed and envied the remarkable hunting "equipment" of wolves: their keen noses, their swift running ability, and their inborn tendency to work with others in a pack, both to bring down their prey and to raise their young. Because humans are incurably fond of taking in other animals, especially cute youngsters—because, as I argue throughout this book, it is an integral part of the human makeup to live with animals—probably someone brought home an orphaned wolf cub or even an entire litter.

As must have occurred more than once, some of the cubs did well in captivity and were readily tamed. The ones showing endearing and useful traits would be fed, nurtured, kept around the settlement, and allowed to reproduce. Other cubs were perhaps too wild, too rambunctious, or too aggressive toward humans. Those would be kicked out of the home or more probably killed and eaten.

Generation after generation, consciously or unconsciously, ancient humans selected the individuals with the most desirable traits and favored their offspring, while they destroyed, neutered, or rejected those who were too fierce, too wild. In time, a population descended from wolves had had its genetic makeup altered by human activities in such a way as to predispose the wolves' descendants to live easily and usefully with humans. With care and careful breeding over generations, the genetic underpinnings of the desirable traits were fixed in the population, which means those genes became extremely common and were likely to be passed on to future generations. The result was an entirely new and domesticated species.

What does it tell us that dogs were the first species to be domesticated?

And what were they domesticated for?

The first dogs probably helped hunters find or track prey and protected their home territory and their social group (which included humans), behaviors that are common among living canids such as wolves, coyotes, or jackals. These were the reasons that dogs were domesticated—other than that irresistible human urge to take in and nurture and feed other species. Dogs were domesticated because the human trait of living with animals and being closely connected to them was part of our genetic and behavioral makeup long before the revolution in food security that Childe called the Neolithic.

Canid scholars Raymond and Lorna Coppinger of Hampshire College maintain that asking why humans domesticated dogs isn't even the right question. They argue that wolves domesticated themselves into dogs by hanging around human villages and settlements.

Raymond Coppinger explains their model:

You develop a village. You get food in the village. Guess who comes? [The wolf.] . . . Far from we humans domesticating them, dogs invaded us. Not as pets, but as pests. . . . It's the rules of natural selection. It's Darwinism if you will. They're coming to the food. They're coming to the waste products and the thing about humans is that there's tremendous numbers of waste products. Scavenging on village wastes was a wonderful strategy for the early wolf-dog— and maybe they weren't such a complete nuisance after all. Rubbish dumps breed disease. A mobile post-Pleistocene garbage service might have come in handy. . . .

Twelve thousand years ago when our ancestors first settled down in stone villages, a new niche was formed— the village dump. It split the wolf family in two. The cautious wolves remained apart. But the curious, friendly ones selected themselves to enter our lives. The result was a new species, the village dog.

In the Coppingers' scenario, the protodogs that were less likely to flee from humans got the most scraps and the kindest treatment, creating a self-reinforcing system of lessened aggression toward humans and an increasing identification of humans as pack members. It is an appealing idea.

Self-domestication is a fascinating concept, which rightly emphasizes that there were two sides to any domestication: the human one and the animal one. Both participants had to cooperate or domestication did not happen. Unfortunately, whether or not dogs self-domesticated is difficult to test. Self-domestication would look just like human domestication in the archeological record. Certainly, many canids have benefited from food sources scavenged from human settlements—coyotes and foxes, to name two—and have become more accustomed to human presence as they scavenge (as have bears, raccoons, and several other species). But then we have to ask why only wolves have self-domesticated. Why not coyotes, foxes, and other canids? There is no obvious answer.

New evidence contradicts the Coppingers' hypothesis by showing that the dog was domesticated long before the appearance of agriculture or the first village.

With no village or even semipermanent settlement with a garbage dump to hang around, could dogs in fact self-domesticate? Could wolf packs have tracked nomadic hunters from place to place, following trails of human refuse? If wolves did this as part of a process of self-domestication, those following packs would inevitably have crossed into territories belonging to other packs. Today, when a wolf pack encroaches on another pack's territory, the residents usually defend their territory vigorously. Some so-called "migratory wolves" have been observed entering the territory of other packs when resources are scarce, but those ventures last days or weeks, not years or decades. Self-domestication simply cannot happen that rapidly. So, how could it happen that one pack—five packs, twenty packs—eventually followed human groups long enough to become domesticated? I don't think they did. The scenario of self-domestication is very hard to envision if people were still wandering seminomadically, and the evidence says they were.

That evidence comes from a remarkable study published in 2009 by an international team of researchers, who found a stunningly early date of $31,680 \pm 250$ years ago for the earliest dog. This is roughly 20,000 years before the generally accepted date of the Neolithic Revolution at 12,000 years ago. Their work is causing a radical rethinking of the Neolithic Revolution and of the process by which domestication may have occurred.

Led by a Belgian expert in Paleolithic carnivores, Mietje Germonpré, the team began its research project by thinking that very early dogs might have been overlooked in the fossil or archeological record. What they needed was a good way to distinguish early dogs from wolves: a technique that could be applied to fossils.

They began by measuring the skull anatomy and teeth of a sample of forty-eight wild wolves and fifty-three modern dogs representing eleven different breeds. They found that modern wolves could be reliably separated from modern dogs by a combination of measurements of the proportions of their snouts and the size of their teeth. Simply put, wolves tend to have larger teeth and longer, narrower snouts than nearly all dogs.

Statistical analysis of the data separated the sample into six natural clusters, with only one outlier: the skull of a modern Central Asian shepherd. One cluster contained all of the modern wolves. Another cluster consisted of modern dogs of a particular type: dogs like chow chows and huskies, which they called dogs of archaic proportions. The third cluster included dogs such as German shepherds and Malinois that have wolflike proportions. These first three groups overlapped with each other to a small extent but were still largely separate. A fourth group of modern dogs with short tooth rows—great Danes, mastiffs, and rottweilers—overlapped slightly with the archaic dogs but not with the others.

The remaining two clusters were entirely separate from all others. One consisted of dogs with extremely long, narrow snouts like Irish wolfhounds. The other included all of the prehistoric dogs, which had long tooth rows, like wolves, but shorter, broader snouts than wolves. These prehistoric dogs didn't overlap with any other group but fell between the modern wolves and modern dogs, suggesting that the prehistoric dogs evolved from a wolflike anatomy.

Finding that prehistoric dogs fell between modern dogs and modern wolves in terms of morphology was no surprise to Germonpré's team. Although geneticists who study dogs and wolves don't always agree, the one point upon which they are unanimous is that dogs evolved from ancient wolves, probably a subspecies of the gray wolf, *Canis lupus*. It is only to be expected, then, that early dogs would have wolflike skulls and teeth.

This initial part of the study by Germonpré's group established which skull measurements were useful in determining what was a wolf and what was a dog. The six clusters formed by their data proved to be statistically very reliable and well separated.

They then added to their database comparable measurements on a series of seventeen "unknown" specimens that they wanted to classify. Seven of these were not truly unknown; they were five young wolves and two zoo wolves. Why include such animals as unknowns? Immature animals often look different from their adult relatives, witness the differences between a kitten's body proportions and an adult cat's. Zoo animals were also chosen because they are often physically abnormal, due to the unnatural conditions under which they are kept. One of the "unknowns" was the modern Central Asian shepherd, which didn't fit into any of the clusters in the first part of the study. Finally, they added a series of eleven skulls of large fossil canids from Belgium, Russia, and the Ukraine, although two of these proved to be too incomplete to be classified. Some of the fossil skulls had been sitting in the museum since the 1860s, when they were first excavated.

The team used a statistical technique known as discriminant function analysis to classify the unknowns into the preexisting six clusters. Satisfyingly, all of the immature wolves were classified as wolves. The two zoo wolves were classified as recent dogs with wolflike snouts, reflecting their odd morphology. Five of the unknown fossil canids slipped easily into the modern wolf group and were concluded to be wolves. Another fossil fit into the group of recent dogs with wolflike snouts—even though it was older than 22,000 years—showing it had both wolflike and doglike features.

The remaining three fossil skulls—one from Goyet Cave in Belgium and one each from the Ukrainian sites of Mezin and Mezhirich—resembled each other closely (Figure 29). Even without considering specific measurements, the team knew these were unusual-looking skulls for prehistoric wolves. In fact, another researcher had already suggested that the Mezin skull was an ancient wolf that had been kept in captivity.

All three of these fossil skulls were placed into the prehistoric dog cluster by the statistical classification technique. An advantage of using discriminant function analysis is that it not only assigns

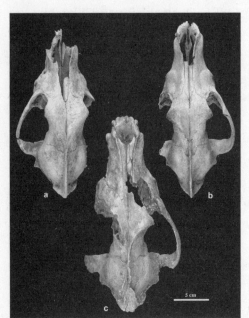

29. *The cranium of the earliest known domestic dog, from Goyet Cave (a), shows different proportions from the crania of two similarly ancient wolves from France (b, c). The dog has a relatively wider snout and larger braincase than the wolves. Wolves also have shorter tooth rows than dogs. Modern wolves show the same proportions as the fossil wolves shown here.*

specimens into preexisting groups, it also gives an estimate of the probability that the classification is correct. The Goyet skull had a 99 percent probability of being a prehistoric dog; the Mezin fossil a 73 percent probability; and the Mezhirich skull a 57 percent probability of belonging in the prehistoric dog cluster.

Ninety-nine percent probability and 73 percent probability of being a dog sound convincing. A 57 percent probability sounds less persuasive until you realize that the next most likely classification (the modern wolf cluster) for this individual has only an 18 percent probability of being correct. Therefore, the chance that this skull is accurately identified as a prehistoric dog is almost three times higher than the chance it belongs to any other group.

The modern Central Asian shepherd also grouped with these three fossil skulls and was classified as a prehistoric dog with a 64 percent probability. The only other group that might have included the shepherd was the recent wolf group, with 32 percent probability. The large, blunt face, small ears, and stocky build of a Central Asian shepherd give it a rather primitive look, which provides intuitive support for its classification as a prehistoric dog. The head of a Central Asian shepherd gives us a glimpse of what the first domestic dog apparently looked like.

Germonpré and her colleagues were delighted with the results. They had hoped to find a way to recognize prehistoric dogs in the fossil record and they had developed one. They had used their technique to identify three new prehistoric dogs. A key question remained: How old were these skulls?

The Mezin and Mezhirich skulls were associated with artifacts of the Epigravettian culture, which occurred between about 14,000 and 10,000 years ago in the Ukraine, where they were found. A more precise dating of the skulls awaits the application of the radiocarbon technique.

The Goyet skull was another matter. Since Germonpré works at the Royal Belgian Institute of Natural Sciences in Brussels, which owns the Goyet Cave collection, she was able to take a sample of the skull for radiocarbon dating. The sample was sent to the Beta-Analytic Laboratory, one of the premiere dating labs in the world. Using accelerated mass spectrometry (AMS) radiocarbon dating—which requires only a few grams of bone—the lab produced a date of $31,680 \pm 250$ years old.

"I was not so surprised at the rich genetic diversity of the fossil wolves," says Germonpré, citing studies by others that have found the diversity of ancient wolves to be higher than the diversity of wolves today. Foxes and wolves underwent a severe bottleneck in population size at the end of the last Ice Age, and many genetic lineages went extinct at this time.

"But I *was* surprised at the antiquity of the Goyet dog," Germonpré adds. "I expected it would probably be Magdalenian." Goyet Cave includes five separate stratigraphic layers preserving artifacts from many different cultural eras. In that cave, the Magdalenian is roughly 18,000–10,000 years old, which would be similar to the dating of early domestic dogs from sites in Russia at 14,000 years ago. Finding the Goyet dog dated to almost 32,000 years ago was astonishing. But the date was calculated by a well-regarded, experienced radiocarbon lab. The date also had a very small error of ± 250 years.

Since the Goyet dog is so old, there must be a large gap in the fossil record of early dogs between 31,700 years ago and 14,000 years ago. Yes, the team expected that some early dogs had been overlooked because there were no good techniques for identifying them. But is it reasonable that almost 18,000 years of early dogs had been missed?

Mietje Germonpré thinks so. First of all, to use the method her team developed, you need a nearly complete skull, and canid skulls are rare in this time period. Other bones can't be used because there is still no reliable way to separate a dog femur or thigh bone from a wolf femur, for example. Second, as she points out, not many scientists have even asked the question whether a large canid skull from such an early period could be a domesticated dog. Such specimens have usually been judged "too ancient" to be domesticated. It is better to approach these problems with an open mind and let hard evidence lead you to the correct conclusion, she believes. Finally, if domestication was a chancy event—and most animal breeding is, even today— there were probably only small populations of domesticated dogs during that time. Probably few of the earliest dogs would have been preserved as fossils and then found by paleontologists.

Germonpré's team also sought evidence to tell them what these early wolves were eating by analyzing stable isotopes in their bones. The food an animal eats enters its bones and leaves a characteristic chemical signature, in the form of carbon and nitrogen isotopes. If the preservation of the bones is good, this chemical signature remains in fossils and reflects the diet during the last few years of an animal's life. Although exact diets cannot be reconstructed, isotope analysis shows whether an animal used marine, freshwater, or terrestrial resources. Stable isotope ratios also reveal where an animal was in the food chain compared to other animals from the same site. Herbivores that eat predominantly terrestrial plants have low nitrogen values, whereas carnivores that feed on those herbivores have much higher levels of nitrogen. Isotope analysis may also reveal broadly which kind of plants were eaten because of differences in the metabolic pathways of plants.

Ten canid bones from the Belgian caves of Goyet, Trou du Frontal, Trou de Nutons, and Trou de Chaleux were analyzed for stable isotopes, along with seventy-two herbivore bones (horse, reindeer, goat, red deer, musk ox, bison, and hare) from the same sites. None of the canid bones sampled were skulls that could be classified, so the team took as their basic assumption that all canid samples were from wolves.

The results showed that these ancient wolves most probably relied heavily on horses and bison as prey, while musk ox, reindeer, hare, and salmon—which occur as part of the human diet at later Upper Paleolithic sites—were not dominant components of the diet. While humans from Goyet were domesticating dogs, they were almost certainly sharing food with protodogs and wolves.

What these findings suggest is that very large prey (horses and bison) were shared and smaller items were not. From a human perspective, this is a sensible approach to the process of domestication. If food is abundant, throw some to the dog or protodog to reinforce a friendly relationship. But if food is scarce and the entire human group is dining on a few hares or a salmon, give little or nothing to the protodog.

Dogs improved success rates in hunting because of their superior skills in scent tracking and their speed in running. They also almost certainly served as alarms, alerting humans when other predators came near. In modern settlements in the far north, one or more dogs are commonly kept staked outside the house. The dogs are fed and cared for, but never brought into the house, for their purpose in life is to warn their humans of the approach of dangerous polar bears by barking. It is easily envisioned that an ancient domestic dog might have similarly warned humans of the approach of bears, wolves, or simply strange humans.

We don't know for sure when dogs became such complete household members that they would be best described as working pets rather than living tools. There are other early dog fossils: one dated to 17,000 years ago from the site of Eliseevichi, Russia, in the Bryansk region; one dated to 14,000 years ago from Oberkassel, Germany; and another dated to about 14,500 years ago at Kesserloch Cave, in northern Switzerland. Certainly, this transformation took millennia if not longer. Finds from 12,000–13,000 years ago in Israel suggest that the change to dog-as-member-of-the-family had been accomplished by the time that an elderly woman was buried at Ein Mallaha, her hand resting on a puppy in a gesture that clearly speaks of affection and intimacy. This burial was well after the domestication of many other species, too: sheep, goats, cattle, and various types of grain, suggesting that the Ain Mallaha dog might have been a useful herding dog.

My old friend Bob Wayne of UCLA has been pursuing the evolutionary and genetic history of various canids since I first met him in his graduate student days. Now his lab at UCLA is widely acknowledged as one of the international leaders in such studies. In 2010, his team published a genomewide examination of a large sample of wild and domestic dogs, looking to trace the ancestry of modern dogs. The extent of their genetic survey and the numbers of individuals involved far surpassed those of any previous studies. They examined the DNA of 912 individual dogs from 85 breeds, as well as that of 225 gray wolves from eleven regions of the world. They were looking for the presence of any of 48,000 single nucleotide polymorphisms (SNPs, pronounced "snips"), which is where one of the four bases that make up DNA has been replaced by another. Each SNP is a very small mutation that records a bit of evolutionary history. By looking at which breeds and wolves share suites of SNPs, the team could trace genetic ancestry.

Bob's group found that domestic dogs consistently carry more genetic input from the Middle Eastern wolf than from the East Asian wolf. Previous studies had suggested that the East Asian wolf was the primary ancestor of modern dogs, an idea which is contradicted by this study. How had earlier studies mistakenly identified China or East Asia as the home of the wolf that gave rise to dogs? Those studies had smaller samples and looked at fewer genetic markers, Bob explains. Probably once dogs were domesticated and fully integrated into human households in the Middle East, they spread into East Asia and interbred with local wolves. The only modern dogs with a high contribution of Chinese wolf genes are genetically unusual breeds—a sign of an ancient split from the other breeds—that persist in Asia today: the Akita, Chinese shar-pei, chow chow, and the dingo.

Dogs were not the only domestic animal, simply the first species that took advantage of new opportunities for communicating with humans.

14

Signs of Domestication

I ARGUED IN the last few chapters that there was no abrupt "human revolution" at the beginning of the Upper Paleolithic period, no sudden leap forward that marked the beginning of modern human behavior, and no Neolithic Revolution at 12,000–10,000 years ago that marked the rise of human life with domesticates either. Rather than envisioning a series of technological revolutions, I see human behavior evolving fairly gradually over the last 100,000 years or so as new behaviors were invented in this population or that population. Logically, it seems certain that a number of useful behaviors were invented independently by separate groups. For example, I suspect many human groups that were genetically capable of inventing tools or language did so—and then died out, simply because their numbers were too low and because random events can be devastating to small populations. However, until the density of humans on the landscape was high enough that different groups encountered each other frequently, none of these behavioral or economic changes spread specieswide. No matter how clever you and your group were, if your idea never reached another similar group, the idea died with you.

If we look in detail at the changes in lifestyle, hunting, art, personal adornment, and toolmaking over the last 100,000 years, we can see that some changes or innovations appeared in one place, then disappeared, and still later reappeared in another.

From the dates associated with the Goyet dog and others, this first episode of domestication—the transformation of a wolf to a dog, changing a dangerous competitor into a friendly helper—happened at almost the same time as the beginning of figurative art and music. But this first instance of domestication did not set off a revolution—or, more accurately, the revolution that spread farming across much of the world didn't happen until about 20,000 years later.

Even though I don't believe there was a Neolithic Revolution *per se*, I do believe that learning to create living tools out of suitable animals eventually altered the adaptive niche of the human species. An idea called "niche construction" is now very popular with biologists, who are seeing with new eyes the many ways in which a species can actually tailor its environment to its own advantage, constructing its own niche, so to speak. In a sense, the Living Tool Revolution was a clear example of niche construction, as were the later and more widely recognized ones like the Industrial Revolution. The Living Tool Revolution enabled humans to harness (literally in some cases) sources of power far beyond our own puny abilities. But whereas the much later Industrial Revolution taught us how to harness mechanical power, which changed the society and economy of Western Europe in a few generations, the Living Tool Revolution took thousands of years to work its changes.

This is because mechanical power is a great deal less individualistic than animal or plant power. Engines—whether steam- or coal- or oil-powered—are much the same and work on the same principles. In contrast, each species of animal that has been domesticated brings its own idiosyncratic characteristics and needs to the process.

In the last chapter, we looked at the first domesticate and how dogs might have been domesticated and changed by living with humans, according to fossil and genetic data. Many different species have been domesticated over time: at least fifteen large-bodied mammals and many smaller mammals or birds (see Table 1). The evidence for some domestication events is excellent; for others it is poor. Surveying the whole record brings a few points into sharp focus.

First, dogs were domesticated well before the other species, a fact which brings into serious question the idea that animals were domesticated as meat sources or walking larders.

Second, the domestication of dogs at 32,000 years ago was followed by a cluster of animal domestications in the Neolithic Revolution period 12,000–10,000 years ago; these were goats, sheep, pigs, and cattle. All these species fit reasonably well into a model in which the farming of domesticated plants—staple crops—was combined with herding.

But the process of domesticating a wild sheep or goat or even wild cattle (aurochs or buffalo) can have been nothing like that of domesticating a wolf. For one thing, the wolf can and will eat a human. Cattle, sheep, and goats may run away, kick, even gore humans, but they aren't predators and don't regard humans as natural prey. Indeed, the Neolithic Revolution species are far more likely to fear humans than eat them, so the

TABLE I—MAJOR DOMESTIC ANIMALS

Animal	Region where domesticated	Approximate date (years ago)
Dog (*Canis familiaris*)	Western Europe	32,000
Goat (*Capra aegagrus hircus*)	Western Asia, Iran	12,000
Sheep (*Ovis orientalis aries*)	Southwestern Asia	11,000
Pig (*Sus scofa domesticus*)	Near East, China, Europe	10,000
Cat (*Felis silvestris catus*)	Near East	9,000
Cow (*Bos taurus primigenius*)	India, Middle East, Africa	8,000– 10,000
Alpaca (*Lama pacos*)	Central Andes	8,000
Llama (*Lama glama*)	Central Andes	6,000
Water buffalo (*Bubalus bubalis*)	India, China	6,000
Horse (*Equus ferus caballus*)	Kazakhstan, Eurasia	6,000
Dromedary camel (*Camelus dromedarius*)	Arabia	6,000
Donkey (*Equus assinus*)	Northeast Africa	5,000– 7,000
Reindeer (*Rangifer tarandus*)	Russia, Finland	5,000
Bactrian camel (*Camelus bactrianus*)	Central Asia	4,500
Yak (*Bos grunniens*)	Tibet	2,500– 5,000

skills and behaviors needed to first tame and then domesticate these later species were quite different from those used to manage wolves and dogs. Still less does domesticating an animal resemble domesticating a plant!

Third, all animal domesticates seem to have been domesticated more than once in more than one geographical location. On first hearing, this idea is surprising because domesticated species are genetically different from their closest wild relatives almost by definition. So, how could those changes happen more than once? The paradox disappears when domestication is viewed as a process through time. At various points in the process, and certainly in the beginning stages, the species undergoing domestication was not yet so distinct from the wild one as to prohibit crossbreeding or back-breeding. Wolves still crossbreed with domestic dogs upon occasion, despite millennia of genetic separation.

Fourth, each domestication event is individual and peculiar, so we need to examine not only the first domesticate but also the last and some intermediate domesticates to understand the full range of the process. Dogs are not sheep, horses, cats, cattle, nor goats, llamas, pigs, or camels. What had to be learned, intuited, negotiated, and established between individual humans and the wild animals with which they were living (sometimes successfully, sometimes not) was different every single time. Those humans who observed more keenly, learned more quickly how to quiet a frightened or anxious animal, and figured out how to keep it alive and docile were rewarded with a new kind of tool: an animal that would do the bidding of humans or at least cooperate with them.

Finally, we need to address a thorny problem I sidestepped in the previous two chapters. In discussing dog domestication,

I described some of the skull features that showed Mietje Germonpré and her team that the Goyet dog was indeed a domesticate. If each instance of domestication is different, then how do you recognize an early domesticate that is not a dog? How can you distinguish between a domestic species and its wild progenitor if all you have is a fossilized bone?

The general principle of domestication is that those attributes humans favored are more strongly developed in a domestic species than in its wild progenitor or relatives. If what humans wanted was a large, succulent fruit, then the fruits of domesticated plant species are bigger. If the aim was to domesticate a swift animal that could transport people and goods rapidly, then you look for a big, fast target animal. The question is whether we can actually discern what our ancestors wanted from a target species, which is sometimes little more than a paleofantasy.

In a review of domestication in 2006, Melinda Zeder of the Smithsonian Institution and her colleagues defined domestication as "a process of increasing mutual dependence between human societies and the plant and animal populations they target. . . ." Both "mutual" and "dependence" are key words here. How is this process recognized? They explain that geneticists look for markers of domestication in living domestic species, while archeologists look for evidence of human behavior patterns that changed the behavior and morphology of domestic species—and, not incidentally, their genomes.

Recognizing that a species has become domesticated in the fossil or archeological record is not easy and is hence often controversial. In 2005, Jean-Denis Vigne of the Centre National de la Recherche Scientifique and his colleagues identified nine clues that might attest to domestic status.

Domestic animals are often portrayed in artwork that shows the species was incorporated into the society's symbolic system. This criterion alone surely does not suffice unless we are willing to accept that all of the animals portrayed in prehistoric art were domesticated, including the woolly mammoths, lions, bears, and rhinos. While depiction in prehistoric art clearly shows that something was very important and symbolic about these species, we know that many of the exquisitely depicted animals in prehistoric art were never domesticated and were probably not able to be domesticated. And yet, dogs were the first domesticate, and their domestication occurred only a few thousand years after the earliest figurative cave art, so you could argue that the depiction of dogs proves the value of looking at artworks.

Countering this interpretation is the fact that dogs appear so rarely in prehistoric art, perhaps fewer than a dozen times. In fact, dogs are even scarcer in early cave art in Europe than are humans themselves. When I asked Paul Bahn, an independent scholar and an expert on prehistoric art, why this should be so, he replied, "Probably for the same reason that human figures were also very rare—either it was taboo to depict them for some reason; or (far more likely) they were simply not relevant to what the artists were mostly doing."

Anne Pike-Tay, a professor at Vassar College who researches the Upper Paleolithic period, made another suggestion. "The scarcity of artistic depictions of carnivores parallels their scarcity in the fossil faunas of the Upper Paleolithic," she observes. If domesticated dogs were helping humans hunt, perhaps they were placed in a completely different symbolic category from other animals. She wonders: "What if dogs were put in the 'human family' category as an extension of the hunter, and like humans, warranted no (or very few) painted or engraved depictions?"

Another sign of domestication is a dramatic change in the representation of various animals in archeological remains over time, showing a dominance of a particular species. This approach can work but can also be problematic. Some groups of people—and therefore some of the archeological sites they leave—have or had specialized patterns of hunting wild animals, and these preferences can strongly bias the bones they leave behind. In modern Pennsylvania, for example, white-tailed deer are the most commonly hunted wild species. Between 323,070 and 504,600 deer were harvested by hunters each year from 2000 until 2009.

What this means in practical terms is impressive. Where I used to live in central Pennsylvania, the mail was not delivered or the trash picked up on the first day of deer season, and the schools were often closed because "everyone" was out getting their deer. In such rural areas, deer bones must make up a significant proportion of the animal bones in the refuse generated by hunting families, even though deer are not domesticated. If hunters in the past concentrated on a single species in the same way, then the bony abundances could misleadingly imply the domestication of that species.

Similarly, when Sandra Olsen was my postdoctoral student in the 1980s (she's now at the Carnegie Museum of Natural History), she was working on the bones from a well-known kill site in Burgundy, France, called Solutré. Between 32,000 and 12,000 years ago, several thousand horses were killed in a natural cul-de-sac on the side of the steep Roche de Solutré, a lime-

stone ridge that rises well above the floodplains of the Saône
River in that region. Of the 5,000 bones Sandra examined, fully
94 percent were those of horse, with a much smaller number
of bones from other species, including reindeer. Most of the
horses were killed in the summer months.

Many of the early interpretations of Solutré, including some
dramatic engravings, suggested the site was a horse jump, anal-
ogous to the famous bison jumps used by Native Americans.
Bison jumps show thousands of bones of the single species
that was driven off a cliff at locations like Head Smashed In in
Alberta, Canada. Alternatively, finding almost 4,500 bones of a
single species at any sort of human living site might suggest that
the species was domesticated and simply slaughtered as needed
for food over a long period of occupation.

Sandra's detailed analysis of the vast accumulations of horse
bones at Solutré shows the bones were deposited over many
years as Paleolithic hunters took advantage of the seasonal
migration routes used by horses. During the warmer months,
small family bands of horses regularly migrated from their winter grazing grounds on the floodplains of the Sâone River to the
highlands of the Massif Central. The topography of the region
dictated that they had to pass through a relatively narrow area
between the Roche de Solutré and the next adjacent limestone
ridge. In all probability, hunters diverted single bands of horses
into the cul-de-sac on the side of the Roche, using drive lanes
constructed of brush. Once the horses were forced up against
the Roche, they could be slaughtered relatively easily. What
this example shows is that specialized hunting techniques, as
well as domestication, can produce large numbers of bones of
a single species.

A technique for identifying domestication that has recently received new attention is a shift in the pattern of age and sex representation in animal remains which suggests efficient patterns of butchery/exploitation. The choices about which domestic animals to slaughter on any particular day of the year determine what goes into the resulting pile of bones. Many animals breed seasonally and give birth to singletons or litters (multiple offspring) at whatever time of year their food is likely to be most plentiful. In temperate climates, this breeding pattern means that offspring are produced in the late winter or early spring, to eat and grow through the spring, summer, and fall. An ideal domestic animal being raised for meat will reach nearly adult size by late fall or winter of its first year, perhaps growing faster than a wild species.

But domestic animals in managed herds die in different patterns from wild animals. In the wild, animal populations are diminished by the deaths of the weakest animals, with mortality peaking during the most difficult season. You expect high mortality among newborns and among old individuals, with prime-aged adults being mostly likely to survive. In contrast, among domesticated and managed herds, the mortality of newborns is relatively small. Instead, many individuals are slaughtered at the age of 6–12 months and at the onset of winter, because they have nearly reached adult size and feeding them over the winter will yield little additional meat. Only those animals needed to breed the following year will be spared: usually the best females and a small proportion of young adult or adult males. The selective harvesting of young males is another telltale sign that herds are being actively managed by humans.

Brian Hesse is a colleague of mine at Penn State University who specializes in zooarcheology (the study of animals in archeological sites). In 1978, he was among the first to realize that early domestication could be detected in changes in the age and sex distribution (mortality profiles) of bones at archeological sites. In 2000, he and Melinda Zeder published a now-classic paper following up on Hesse's work on mortality profiles. They showed, first, that goat bones of males could be separated from those of females by their size, and, second, that if you looked at the age at death of male and female goats separately, you could see if young males were being selectively harvested. Since managing the sex ratios of a herd of wild goats sounds impossibly difficult, managed herds must be domesticated herds. Using this approach, they were able to prove that domestic goats were present at a 10,000-year-old archeological site, Ganj Dareh, in the Zagros Mountains of Iran, which is the earliest known evidence of domestic goats.

Domesticated wheat occurs at about 10,500 years ago at a Turkish site called Nevali Çori, and domesticated rye was found at Abu Hureyra in Syria at 13,000 years ago. Combined with the work on goat domestication in the Zagros Mountains, this seems solid evidence that Childe's classic Neolithic package of the domestication of some animals and plants was in place by 10,000 years ago, and possibly earlier. Sheep and goats—sometimes called "small cattle" in historic texts—fit nicely into a lifestyle that also involves farming grains. Grain growers need to live in fixed or at least semipermanent settlements in order to tend the fields. Sheep and goats can be taken to pastures away from the settlement during growing season, either on a daily basis, or simply driven up into the mountains and highlands that were their ancestral home. There, the animals can roam and feed for several months with only a few humans to protect them from predators until the crops are harvested, then they can be driven back to the lowlands to feed on the stubble in the fields.

For herders of sheep and goats, herding dogs offer a huge advantage. A man or woman with a good herding dog can easily manage a few thousand of these animals; a person on his or her own can manage far fewer. Were the dogs living in human settlements at about 12,000 years ago herding dogs? It is difficult to tell. Most modern breeds have been seen as a product of the Victorian era, when pedigree keeping and deliberate selection for dogs of different appearance or function became popular.

However, the recent genetic study of modern dog breeds by Bob Wayne's group at UCLA suggests that the separation of dogs into functional groups might be much older than anyone has thought. The team was surprised to see that different functional groups of breeds (herding dogs, working dogs, companion dogs, guard dogs, and so on) formed distinct genetic clusters. While the team was working up the data, graduate student Bridgett vonHoldt was responsible for color-coding each individual on the breed's function. When Bob first looked at the color-coded results, he thought he hadn't explained clearly to Bridgett what he wanted her to do, since she seemed to have color-coded the results by genetic groupings, not functional groups.

"Then I realized the two were nearly identical," he says. "We had expected there would have been lots of ways to build a sight hound, or a herding dog or terrier, for example." He believes the close correspondence between the genetic data and information on breed function reflects how people set out to make a new breed with a particular function.

"Apparently if people wanted a new sight hound," explains

Bob, "they tended to cross sight hounds with each other, and the same with herding dogs and retrieving dogs. That may not seem so surprising, but we had no reason to think beforehand that these groups would be strongly genealogical." The rich connection between dogs and humans—between dogs as living tools and humans as tool designers—developed and changed over a long, long period.

One very different criterion of domestication is a change in the shape of the horn core of sheep and goats. Something about the traits human selected for as they were domesticating sheep and goat—perhaps less aggressive behavior—caused the change of horns in domesticates or caused horns to be lost altogether among females. Such a change seems to be a strong indicator of domestication.

Another convincing criterion is the abrupt appearance of a species in an area where it and its close relatives were previously unknown. As a general rule, people don't transport large wild animals to new areas because of the difficulty of managing them. Nonetheless, examples of the deliberate introduction of wild animals to new areas are known. For example, in the 1850s, red deer were transported to New Zealand and Australia, as was the European rabbit. In 1896, moose were introduced to Anticosti Island off Canada from the mainland. Various types of deer are among the most commonly transported and introduced animals; although no one has ever succeeded in domesticating deer, they are relatively docile (compared to, say, rhinoceros, wild horses, or predatory species). Wild boar—or perhaps semi-domesticated pigs—were apparently transported by humans to various small islands off Japan, where their bones are found in archeological middens. This means that, to use the sudden appearance of a species in a new area as a criterion, you need to think about the animal's behavior.

Many animals get smaller during domestication. This means that a difference in bone size or body size, or in brain size relative to body size, or in the length of the tooth row relative to jaw size (dental crowding), can be a useful tipoff.

Smaller does not equal domesticated in all cases. Humans may select for smaller-bodied animals if meat yield is the concern, because smaller animals tend to mature more quickly and be less aggressive. Pigs are a good example of a domesticated species that diminishes in size at the time of domestication, though later breeding has made modern pigs much bigger. It may be pertinent that many domesticated pigs were spread around the Pacific by people in small boats, a situation in which a large or aggressive animal would be a lot of trouble. However, animals domesticated primarily for power (hauling loads, plowing) or transport don't always become smaller because size is a desirable attribute. Camelids like llama and alpaca seem to get larger with domestication, not smaller, and are also larger in colder regions. Horses seem to get smaller—or at least less stocky and more slender-legged—but their size is also affected by habitat.

An abrupt change in diet (usually revealed by the analysis of the stable isotopes that have been incorporated into an animal's bones from its food) can also indicate domestication. Once humans start controlling a species closely—once it is domesticated or well on the way toward domestication—they often confine the animals. Free-ranging animals will breed with whatever individual they choose, which means that humans cannot influence the traits in the offspring. But confining animals so you can control breeding comes with a cost; in most cases, a confined animal must be provided with food during the hard winter months.

In 2002, Yuan Jing of the Institute of Archeology of the Chinese Academy of Social Sciences in Beijing and Rowan Flad of Harvard University found that the appearance of domesticated pigs in China came after the domestication of foxtail millet as a major crop. They speculated that successful growing of a staple crop might have been a prerequisite to domesticating animals. At the 8,000-year-old site of Cishan, for example, the remains of foxtail millet were found in a series of rectangular storage pits that were so large and numerous that an estimated 111,000 lbs (50,000 kg) of millet were being stored. Clearly, the people of Cishan had a reliable staple crop and enough excess that they could store very large quantities of grain. Yuan and Flad's analysis of pig bones at another site, Xiangfen City, reveals that those pigs were fed on plants with a particular metabolic pathway known as C4 that uses a particular isotope of carbon; foxtail millet is one such plant. The researchers concluded that the chaff from the crops was probably fed to the pigs on a regular basis.

Though DNA can only rarely be extracted from archeological bones, which stymies attempts to recognize domestication on a genetic basis, genetic studies can reveal useful facts about the evolutionary branching of domestic breeds. Genetic comparisons among living pigs (wild and domestic), for example, tell a more complicated story than anyone expected.

The earliest domesticated pigs all turn up around 9,000 years ago, but not only in China. Recent studies by Gregor Larson of the University of Durham and his colleagues show that domestication occurred independently in at least seven regions: Central

Europe, Italy, northern India, Southeast Asia, and maybe even Island Southeast Asia. Even though the farming of domesticated crops began in the Near East and spread outward into Europe, as the Neolithic lifestyle spread, something unexpected happened. Near Eastern pigs were taken to Europe, presumably by migrating Near Eastern farmers, but those pigs didn't last long in Europe. After five hundred years, they had been almost entirely replaced by pigs descended from European wild boars. The researchers suggest that Near Eastern farmers brought the idea of domesticating pigs and the techniques for managing pigs with them, but something about domesticating the local wild boars was more appealing than adopting the Near Eastern pigs. Perhaps the Near Eastern immigrants kept a tight monopoly on their pigs. Perhaps European wild boars were nicer pigs. Who knows?

The remaining criterion for recognizing domestication relies on ancillary evidence: pens, corrals, or equipment for animal handling and changes in genetic composition from the ancestral condition. The existence of equipment related to animal handling or animal use strongly implies that humans were living intimately with animals and to some extent controlling their behavior. Although wild animals can be confined or used, the sheer volume and diversity of such equipment would seem to be excellent evidence of domestication. Nowhere is the power of this criterion clearer than in work carried out by Sandra Olsen and her colleagues on the domestication of the horse.

For years, Sandra and her team have been excavating and studying horse remains from a large archeological site in Kazakhstan known as Botai, and some other nearby villages made by people with the same culture. The Botai culture is dated to 3,700–3,100 years BC. There is an abundance of preserved bones from Botai—nobody has arrived at a definitive number, but Sandra estimates that 300,000 fragments is a reasonable guess—of which roughly 99 percent are those of horses. Many bear butchery marks and cutmarks from being processed with tools. In addition, there are distinctive, deliberate burials of stallions—sometimes whole, sometimes just the head—which shows that horses had an important place in the culture and ritual of the Botai people.

But were Botai horses domesticated or simply hunted and perhaps tamed? The only way to find out was to carry out a lot of detailed studies.

Because male and female horses do not differ strongly in body size, the only way to look at sex ratios is to look at their jaws (males have bigger canine teeth, but these teeth erupt from the gum only in adulthood). Studying all jaws and isolated teeth allowed Sandra to discover that 54 percent of the individual horses from Botai were male.

What does this pattern mean? In the wild, equines generally form two sorts of social groups: bachelor herds comprised of a few males, and family groups comprised of a stallion, his harem of mares, and their young offspring. Smart horse hunters varied their technique according to the number of hunters.

"A communal hunting strategy would normally focus on the family bands because they contain more individuals and are more cohesive than bachelor groups," explains Sandra. "This kind of hunt would leave bones of a high percentage of juveniles— whose sex couldn't be determined—plus an overwhelming pre-dominance of females and a single male: the stallion."

Unfortunately, a managed herd of sheep or goats looks just the same, with many females, lots of juveniles, and few males.

The big difference between managed herds of sheep or goats and horses lies in the practice of gelding (castrating) male horses and keeping them alive. Geldings are more docile than stallions and less unpredictable in their behavior, but they are still very powerful and useful for riding or hauling goods. The fact that the Botai remains contain 54 percent adult males strongly suggests that many of them were geldings. A similar proportion of males has been documented in more recent (2,500–2,920 years ago) managed herds of domestic horses.

Another key point that emerged from Sandra Olsen's studies was that most Botai horses died as mature adults between the ages of five and eight years. Natural mortality produces many deaths among the very young and many among the very old, with few individuals dying in their prime. She argues that this distinctive pattern of prime adult deaths reflects the fact that horses were valued for attributes other than carrying a lot of edible meat on their skeletons. As meat, an adult horse can only be used once, after death; but the same meat is also muscle that can be used over and over again while the animal is alive, for hauling, transporting people and goods, and eventually plowing. A predominance of deaths among mature adults suggests that horses were killed as they began to age and be less useful in work or in producing foals and milk.

At this point, Sandra and the team working at Botai and the nearby contemporary villages of Krasni Yar and Vasilkovka felt they had proved beyond reasonable doubt that the Botai horses were domesticated. Their skeptics still argued that they could be documenting a pattern of deaths based on preferential hunting of bachelor herds of horses.

Sandra's response to the argument pointed to the fact that

entire horse skeletons—all body parts—were found in the excavations. "I doubt that Botai hunters ventured miles from home
overland on foot, killed a horse, and dragged the whole one-
thousand-pound carcass back to the village," she says, citing the
"schlep effect."

Derived from the Yiddish word *schlep* meaning "to carry
awkwardly," this term is used by archeologists to explain why
hunters often strip the meat from the heavier bones at the kill
site and carry home only the meat and smaller bones. Schlep-
ping a whole horse for several miles would indeed have been an
arduous and awkward endeavor. Finding whole horse skeletons
in a village strongly suggests that the village is where the animals were killed. Besides, with 160 houses at Botai, the settlement was sizable. How long could so many people continue
to intensively hunt wild horses without running out of game?
Those horses must have been domesticated.

But, the skeptics countered, where was the evidence of
tack—the bits, bridles, lead ropes, stirrups, or saddles needed
to manage horses? The problem was that the Botai people had
no bronze or iron technology, so the tools of the horse-keeping
trade (if they existed) were made out of perishable wood, bone,
or leather. However, the Botai bones did include 270 bones that
had clearly been modified and used, most in the manufacture
and smoothing of leather thongs. A great many items used in
working horses can be made of leather thongs, but so can items
for other uses. A small number of Botai bone tools—thirteen—
were harpoons that may well have been used in hunting animals,
along with stone projectile points hafted on wooden arrows.

How could nonmetallic items of horse tack be detected? In
1986, David Anthony and his wife Dorcas Brown of Hartwick

College in Oneonta, New York, had suggested that wear caused by using a bit to control a horse left permanent, recognizable damage on the teeth of the lower jaw of horses. Bits are placed in the diastema, the space between the incisor teeth and the first premolar (confusingly called the second premolar in anatomy texts), so mouthing or chewing the bit would in time wear a beveled edge on that premolar (Figure 30). Thus, bit wear could be used as evidence of domestication and riding even in the absence of tack itself, they argued.

Initially, this approach was controversial, since many types of events can cause damage to teeth both during and after an animal's death. Besides, only well-preserved teeth can be measured for bit wear. Recently, experimental studies by Anthony and Brown of modern horses and the effects of putting different types of bits in their mouths (or not, in the controls) has proved their procedures for recognizing bit wear and led to greater acceptance of this approach.

Sandra Olsen's team found that five out of fifteen jaws from Botai which had the correct teeth preserved also showed evidence of bit

Top or occlusal view

Side or lingual view

Unworn premolar

Top or occlusal view

bevel

Side or lingual view

Beveled premolar

|1 cm|

30. (top) The premolar of a horse normally appears roughly rectangular when viewed from the side. (bottom) In horses that wear bits, the action of the bit against the premolar wears a beveled edge onto the tooth, which can be used as evidence of domestication. Rectangles drawn on the occlusal view indicate the area that is altered through bit wear.

wear. Were some horses domesticated and others not? Such a mixture occurs today in many parts of the world where wild and domestic animals interbreed freely.

The critics remained unconvinced. Sandra's answer to the skepticism was to think up new ways to test her hypothesis. Because Botai people lived in villages, her team was able to use post molds—discolorations in the soil that show where posts had been dug in to support structures—and a technique of remote sensing called magnetic gradient imaging to reconstruct a schematic plan of the settlement (Figure 31). The wooden houses were dug into the ground, with walls of adobe and roofs of saplings, covered with clay and manure. The houses are rectangular and arranged in rows along streets and around small plazas. There were 160 houses in Botai, 54 in Krasni Yar, and 44 in Vasilkovka. The team also found post molds of large circular or semicircular enclosures in the settlements. Could these large enclosures be corrals or paddocks?

They invented a new and creative way to find out: chemical analysis. Soil samples taken from within the large enclosures showed much higher levels of phosphorus—a mineral found in horse manure—and salt—found in urine—than for soil samples outside these enclosures or away from the village. This finding was compelling evidence that horses were managed and kept in enclosures, an approach that works much better with domesticated horses than with wild ones. Though phosphorus and salt could be deposited by other animals, there were no remains of sheep, goats, or cattle at Botai sites. Remember, 99 percent of the bones were those of horses at the largest site, indicating Botai people specialized in horses whether those horses were hunted or domesticated.

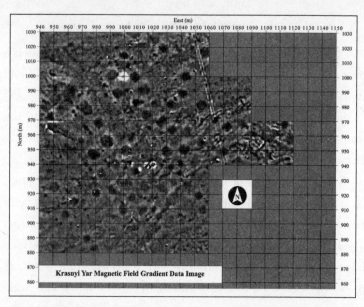

Krasnyi Yar Magnetic Field Gradient Data Image

31. *(above) This remote sensing image of the Botai village of Krasni Yar reveals the locations of postholes and other ancient structures. (below) With the help of excavation data, this remote sensing image is reconstructed as showing rows of houses and roughly circular paddocks for keeping horses.*

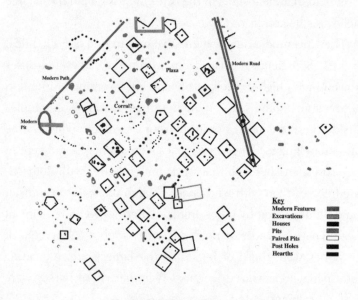

Krasni Yar Village Plan

Though you wouldn't expect animals domesticated for their power to become smaller in overall body size, there is evidence that domesticated horses have slenderer and less bulky bones in their lower legs. Measurements of the length and breadth of metapodials—the bones just above the hoof—show that these bones had statistically different proportions in domesticated horses from the Kent Bronze Age settlement in the United Kingdom (about 1,300–900 BC) compared to much earlier wild horses in Kazakhstan. The proportions seen in the Kent bones—chosen to represent clearly domesticated horses in pre-modern times—are matched by those from a sample of modern Mongol horses and by those of horse bones from Botai settlements. In other words, the Botai people were apparently controlling horses well enough to engage in selective breeding for desirable traits, perhaps greater speed.

But the final piece of evidence that silenced even the most skeptical critics was finding traces of degraded fats from horse milk on the remains of potsherds from Botai. Exacting analysis by Alan Outram of Exeter University and Natalie Shear of the University of Bristol, two members of Sandra Olsen's team, showed chemical evidence that these pots had contained mare's milk. Today, fermented mare's milk (koumiss) is a staple item in the diet of Kazakh horse people, though most non-Kazakhs find it an acquired taste. Finding the milk traces in the pots was the "smoking gun" convincing even the skeptics that horses at Botai were domesticated because, as Sandra says, laughing, "You can't imagine milking a wild mare."

Convincing even the most fervent skeptics that horses were domesticated at least 1,000 years earlier than was previously thought wasn't easy, but it shows the power of using multiple lines of evidence to answer a question. This offered a new perspective on the development of several European cultures.

As Alan Outram explained to a reporter: "What's really key here is that the horses weren't just domesticated. By this point the Botai people have really got the full pastoral package: they were eating horses, they were riding them, they were milking them, which suggests that the original domestication is even earlier still."

Domestic horses revolutionized hunting, transport, communication, food production, and—tellingly—warfare. The entire development of cultures in the steppe regions was transformed by having such an animal. Horses enabled people to live in permanent, year-round settlements because horses—unlike cattle, sheep, or goats—can graze through the snow and live through the harsh winters.

Understanding the impact that domesticating horses might have had on the Botai people and others of the Eurasia steppe is, for me, a review of my own experiences with horses. Even with fully domesticated horses, training them and working with them is a constant exercise in both listening and negotiating. No one can force a horse to do anything (except possibly by using extreme brutality) because they are so much bigger and stronger than we are. Even fences and stalls are really mutually agreed-upon conventions, not absolute controls imposed by humans. And anyone who has learned to ride has, sooner or later, come face to face with the realization that riding means surrendering your physical well-being to an individual of another species. If you can't communicate—if you can't tell what upsets the horse or frightens it or makes it feel cooperative and safe, if you can't make respectful requests of the horse that are understood and accepted—you are very likely to be seriously hurt.

Mounting a horse means voluntarily cooperating in being kidnapped by a creature who may not be even inconvenienced by something that puts your very life at risk. In today's world, when horseback riding is a sport and a pastime, not a necessity, people either learn to pay close attention to communicating with the horse and improve their communication skills or they stop riding altogether, usually after receiving a concussion or broken bones.

People who don't know horses think you simply get on and sit there while the horse does all the work. Riding a horse is a great deal more complex than that. Ill-trained horses may decide to do something other than what is requested of them by a foolish rider. Even well-trained horses—and sometimes especially well-trained horses—may become so annoyed by inept riders that they deliberately try to dislodge them. Relatively few riders are killed today, thanks to things like riding helmets, but just about every rider I have ever known has had several horse-related injuries. The sheer joy of riding horses and communicating with them, and the overwhelming beauty of horses and their movement, become addictive. These factors draw true horse-people back, even though they appreciate the danger of working with animals that often weigh more than 1,000 lbs.

If modern humans getting on thoroughly domestic horses experience these fears, how much more intense would the selective pressure to succeed in horse domestication have been on those humans who domesticated the first horses? Surely their observational and communication skills were stretched to the limit, but the benefit was great and worthwhile. Imagining the process of domesticating horses—using as a basis the varied types of evidence cited here—shows very clearly how strongly domestication affected human society and human behaviors. Not only horses were domesticated in the process; humans were genetically and behaviorally changed to be more domestic, too.

15

Baa, Baa, Black Sheep, Have You Any Wool?

IN THE PREVIOUS CHAPTER, the examples of how domestication can be detected in the archeological record throw into question a number of well-accepted ideas. Summarizing how the animal connection played its role in the more recent stages of human behavior will make this clear.

The facts do not support V. Gordon Childe's notion that there was a Neolithic Revolution—a rapid event—at all, much less one triggered by the first domestication of plants. There was indeed a very important economic shift from hunting-and-gathering to farming and herding, which depended in part upon the domestication of animals and plants, but it certainly wasn't triggered by the onset of domestication.

Between the first domestication (of the dog) at 32,000 years ago and the domestication of the first plant about 13,000 years ago, there was a long time gap. Because Childe implied that domestication was a single concept, this fact would have surprised him. Not me. Why shouldn't there be a gap between animal domestication and plant domestication? The two processes are completely different, in my view, and require different types of skills and knowledge from the domesticators. Then there was another, shorter gap of several thousand years between the appearance of domestic staple crops and the appearance of the domestic animals (sheep and goats) that were apparently first used in the Neolithic mixed-farming lifestyle by about 10,000 years ago. Only when animals suitable for use as domestic stock could be kept in herds and combined with farming of plant staples did the Neolithic lifestyle appear. The Neolithic was not a revolution.

The underlying concept is that Childe's Neolithic Revolution was motivated by a striving for greater food security—knowing where your next meal was coming from—but modern evidence undermines this idea, too. Dogs don't fit the herder-farmer paradigm well at all and they were not domesticated as meat animals, though extraneous dogs were probably eaten. The same sort of discrepancy is shown by the record of domestication of another species I have talked about a lot here: the horse. Based on the findings from the Botai villages in Kazakhstan, horses were more important as a source of power and milk than they were as meat.

Most of all, as Jared Diamond points out in a review in *Nature* of the evolution of domestication, "Food production could not possibly have arisen through a conscious decision, because the world's first farmers had around them no model of farming to observe."

What's more, the payoff from domesticating animals and plants was not what you might suppose. Careful comparisons of the skeletons of Native Americans before and after they took up maize farming show that farming required harder work than hunting and resulted in poorer nutritional status—being dependent on one main crop makes you very vulnerable to crop failure—smaller stature, and more disease, caused by living in dense settlements where disease can spread rapidly.

At this point, you may find yourself surprised by the notion that domestication—or at least early domestication—had serious negative consequences. Why would anyone keep animals or raise plants in that case?

The answer may be precisely that domestication was accidental, not intentional. Diamond argues that plant domestication began as wild plants were carried back to some sort of camp or settlement, where they might (just might) flourish, and that animal domestication arose from "the ubiquitous tendency of all peoples to try to tame or manage wild animals (including such unlikely candidates as ospreys, hyenas, and grizzly bears)." This old idea was first expressed by Francis Galton, Charles Darwin's cousin, in 1865. What Galton and Diamond (and others) noticed was the universality of the human impulse to take in and take on animals; what they both have missed is the evidence showing how ancient the underlying connection between humans and animals is. Without reviewing the long haul of human evolution, as I am doing in this book, you cannot understand that there was an ancient trajectory of biology and behavior that ties humans and animals together in a relationship of increasing intimacy.

Diamond proposes that the engine driving domestication was complex, and he is certainly correct in that. At the end of the Pleistocene era, the world's climate became increasingly unpredictable, turning glacial after a warmer period. The Younger Dryas, as this time is known, is dated to between 12,900 and

11,500 years ago. Abrupt climate change and increasing numbers of human hunters caused a drop in the populations of big game. Humans found that their old ways of surviving were not good enough. They responded by migrating, exploring new habitats and ecosystems, and sometimes expanding the range of wild foods they relied upon to include smaller species, new kinds of species (birds and fish), and plants that required a lot of preparation time. This phenomenon is referred to as developing a "broad spectrum" resource base. There is solid archeological evidence that all of these changes began to occur around 13,000 years ago in many—but not all—parts of the world.

But consider: If what was needed in the short term was more food, and more reliable food, would the domestication of plants or animals have been of much use? Not unless the process was already well underway.

The problem is the demand of the future that domestication brings with it, the paradox that David Webster recognized. You can't eat all the seed or the best seed from your harvest because a portion of it has to be retained and stored for planting the next year. To succeed, you have to switch your strategy from thinking of immediate hunger to planning for future demand. In the same way, many domestic animals must be kept alive and looked after—even fed—if there are to be births the next year.

Traditionally, the well-entrenched concept of a "Secondary Products Revolution"—made popular by Andrew Sherrat of Sheffield University starting in 1981—has provided one explanation for why animal domestication began. In Sherrat's view, domestication was "about" the exploitation of primary products (mostly meat), and only later did people develop a greater dependence upon secondary products (milk, manure, power, wool, and so on) of domestic animals. He proposed that the Secondary Products Revolution around 3,300 BC enabled Neolithic farmers to expand into more marginal areas of the Eurasian continent successfully, leading to dramatic changes in subsistence, economy, and politics that swept across the Near East and Europe in the Chalcolithic and Bronze ages. According to Sherrat, the intensification of the use of secondary products made the rise of complex societies possible.

Haskell Greenfield of the University of Manitoba, who has recently revisited Sherrat's proposed revolution, sees this change as a shift in emphasis from dependence upon primary products to more use of secondary products, rather than as evidence of the first use of secondary products.

Still, the key question underlying this interpretation is why domestication happened at all. If hunting and early domestication yield the same so-called primary products, why bother with domestication?

Brian Hesse, my colleague at Penn State, observes that, in the long run, domestication of animals may yield more meat than hunting, but in its early stages, domestication will not do this. "Successful husbandry requires that animals be withheld from slaughter," he explains, "to provide a basis for starting a flock. Said bluntly, it makes no sense for a group of people, *all* of whom desperately need food, to husband what little they have."

As soon as I read Brian's words, I knew he was right. The primary short-term gain from domesticating animals was not meat but the whole package of renewable resources that animals can offer before they become meat, which Sherrat has mistakenly labeled "secondary products."

Brian and I agree that the beginning of animal domestication was not fueled by the need for the "primary products"— meat, hide, and bone—that dead animals provide. A hunter kills an animal and eats it, perhaps saving a little (if there is some mechanism for doing so, like smoking or freezing meat) for another day. In contrast, a herder puts a lot of time and energy into keeping animals alive and selectively breeding a herd with the qualities he desires. Yet when he kills an animal, he gets nothing more than what a hunter gets from killing a wild animal.

Why bother?

I believe people bothered because, by keeping animals for a while and delaying slaughtering, they could gain a wealth of renewable resources that the hunter never gets: milk, wool, horn, manure, protection, power (the ability to pull, to haul, or to carry heavy loads), and more animals. These products cannot be easily obtained from a wild animal or a dead one. Therefore, I suggest that Sherrat's "secondary products" were actually the primary motivation for the domestication of animals. A herd of cattle is literally self-reproducing wealth of several kinds, whereas a dead buffalo is only a meal.

My interpretation is supported by the archeological record as we know it today. Clearly, the first animals to be domesticated were not food animals and their "secondary products" yielded more economic benefits that were new, previously unexploited or unavailable than meat. Meat from domestic animals is certainly eaten, and always has been, but meat provides no impetus for domestication in the first place. Ask yourself this: If domestication takes 1,000 years (and probably it takes much longer than that), who plans 1,000 years ahead for dinner?

Domestication only provides an advantage if the so-called

"secondary products" came first, once animals were tame enough (or semidomesticated) to allow use of those products. It was Sherrat's "secondary products" that made domesticating animals advantageous.

Another way of looking at the record of domestication is that its real benefit was the transformation of animals into living tools. When you look closely at what a domestic animal can and does provide, there are eleven different abilities or resources that would have given an adaptive advantage to those who domesticated and kept animals.

Domestic animals provide power or traction for transport of people and goods, for plowing, for other tasks well beyond the strength of any human. Domestic animals also offer the possibility of transporting large numbers of people or goods very rapidly over long distances.

Wool or fur can readily be obtained from animals without killing them, and the animals will obligingly grow more. The value of wool for making cloth or twine is immeasurable.

Domestic animals also provide manure, which is an excellent fertilizer, can be burned for fuel and makes a good building material. Yak manure was crucial in settling highland Asia. And every day, domestic animals make more of it.

Domestic animals—at least some of them—offer a cheap means of disposing of food refuse and ordure. Goats eat stubble in the field; pigs and dogs eat excrement and human leftovers. In exchange for this bounty, the animals produce more goats and more pigs and more dogs.

Animals also offer a means of movable storage for excess grain crops. Like the abundant foxtail millet in ancient China, excess crops can be fed to animals. This prevents grain from spoiling or being eaten by rodents, particularly in the time before the domestication of cats. The storage container—the domestic animal—can then be kept, moved, and eaten when needed.

Large, docile animals offer a nutritious new food, milk, for adult humans and, more important, for weanlings. Breast-feeding often lowers a woman's fertility, so if a human baby can be weaned onto animal milk, a mother becomes fertile in a shorter time after giving birth. Thus one woman can have more children during her lifetime and those children can be more closely spaced, without risking malnutrition caused by poor milk production during infancy. Good weaning foods often lead to an increase in the human population by increasing overall productivity and lessening weaning deaths.

Other types of domestic animals provide enhanced protection for people, dwellings, stored crops, and other livestock. Dogs and cats are the obvious examples, but herders have recently started touting llamas as guardians for flocks of sheep.

The domesticated carnivores also provide important assistance in hunting. Dogs are better trackers than humans; they are faster runners, take larger prey, and will hunt with humans. Cats hunt solitarily and are far superior to humans at catching rodents that can decimate crops or carry disease. Dogs hunt with you; cats hunt for you; but both offer an advantage.

Domestic animals also offer a form of mobile wealth that is readily transported in times of war or ecological disaster. A field of maize, for example, cannot be picked up and moved as an army comes through, but animals can be herded away to another place.

Finally, domestic animals offer a combination of traits that permitted people with animals to live successfully in habitats that were previously inaccessible. Without camels' backs for transport, their strength and ability to thrive with little water, their fur for making rugs and cloths, and ultimately their meat, could the tribes of the Arabian Peninsula and the Sahara possibly live in the desert? Without pigs, could the peoples of Oceania have spread from one far-flung island to the next and survived? Could the Inuit thrive in the far north without dogs for transport, warmth, and protection? Could people have moved onto the Tibetan plateau without yaks, or established great civiliazations in the Andes without llamas and alpacas? Clearly not.

In short, domestic animals are really another kind of *extrasomatic adaptation* or tool used by humans to expand the resources they can exploit and the abilities they can call upon. Learning to make tools out of stone requires mastering information about the material properties of different types of rock, their distribution on the landscape, and techniques for producing sharp edges or desired shapes. Learning to make living tools out of animals requires a much more complex and subtle information base.

To keep and domesticate animals requires many kinds of detailed knowledge. You have to know about the animals' biology, ecology, physiology, temperament, and intelligence. For example, if you try to keep pigs in dry, sunlit environments, they will probably die. Pigs are poor at regulating their own temperature and they get sunburned, too. If you don't know that pigs need shade, damp wallowing holes, and lots of water, you could kill a lot of pigs by mistake. They are also extremely good at getting out of enclosures of all kinds to go root up your vegetable garden, so being careful about how your pigs are confined is another important consideration.

To domesticate an animal, you also need some kind of awareness of the principles of breeding for desirable traits. That means you need to understand that the offspring is a blend or combination of traits of the male and the female, even if you don't know a thing about chromosomes and genetics. Using this perception of the principles of selective breeding may be as simple as crossing a docile, cooperative female camel that is a little spindly in the legs with a robust, stoutly built male with a feistier temperament in hopes of getting a calf with sturdy legs but a gentle temperament.

During the process of domesticating a species, you also need to develop an ability to "read" the nonverbal communications of potential or actual domesticates and learn how to communicate with them. This means being able to recognize that the goat's coat is looking mangy or poor or that a horse seems to be unusually aggressive with others. Being able to pick up on these signals, whether behavioral or physical, is important to preventing a small issue from developing into a life-threatening problem.

And if the animals are to work for you—work *with* you—at many different tasks, you need to know how to handle that animal so it is willing to cooperate. You not only need to understand the animal, it needs to understand you. Highly developed communication skills are essential. Even a relatively small domestic animal, like a sheep or goat, can be formidably stubborn, intractable, and possibly dangerous. As for larger and more powerful animals, trying to force them to do something against their will is a sure recipe for injury. An animal keeper must learn how to elicit cooperation, how to request that an animal do something, rather than believing in some fantasy of innate human superiority.

Such skills comprise many aspects of the animal connection. The intensive observation of and intimacy with wild animals—the earlier stages of the human-animal link—were essential prerequisites to acquiring the information base needed to domesticate and use animals.

Success in taming, keeping, and eventually domesticating an animal species was certainly determined by the attributes of the target species. Many who have written about animal domestication have remarkably similar views of the attributes of animals suitable for domestication. Appropriate species need to have hardy young that are able to survive if separated from their mother. They are usually social animals with a natural dominance hierarchy into which humans can insert themselves. Prime targets for domestication must be tolerant of living in close proximity with other members of their species, not highly territorial or aggressive. They need to be willing to breed in captivity or confined quarters and unlikely to panic in enclosures. Finally, they need to be fairly submissive toward humans and versatile in their dietary requirements.

These attributes aside, there is another key factor in the success of any attempted domestication, and that is the skill of the particular humans involved. Especially pertinent were skills in observing animals, in understanding nonverbal communication, and in communicating verbally or nonverbally with animals. Logically, these skills were acquired because of the animal-intensive focus and collection of information about animals that provided a selective advantage to humans during the first two stages of our evolution described here. I think it is likely—but I can't prove it—that there is a genetic component to these skills which might be closely likened to linguistic ability and interpersonal skills.

Because humans who were more successful at handling and living with animals accrued a selective advantage, the animals undergoing domestication were effectively selecting for particular traits in humans. For example, if humans favored less aggressive protodogs, the protodogs would have simultaneously selected for humans with greater ability to read canid behavior for subtle signs of impending aggression and territorial defense.

The domestication of animals altered humans in other ways. An example is the continued functioning of lactase into adulthood in some populations, giving humans the ability to eat milk products into adulthood without gastric distress. The mutations that keep lactase functioning arose at least four times among cattle-herding people in the last 10,000 years.

Other readily observable examples involve using various domestic animals to exploit resources in geographic regions that were formerly inhospitable. The Sami of Lapland organize their lives, movements, and settlements around the needs of the reindeer, not vice versa. Sami are considered reindeer people, and they have shaped their traditions and economy around their animals rather than fitting reindeer into their preexisting socioeconomic and settlement patterns. Similarly, Masai and other cattle peoples inhabit dry regions with sparse vegetation and, like the Sami, migrate as dictated by the needs of their livestock.

The skills and knowledge involved in domesticating, planting, maintaining, and harvesting plants are different from those needed to domesticate, maintain, train, and use animals. Plants don't have to be persuaded to cooperate with humans or to allow themselves to be handled to "give up" their fruits or roots. Plants don't consciously attack people, whether they are domesticated or not, and they don't fail to thrive if they are in a fenced enclosure. Of course, domesticating plants depends upon recognizing a suitable habitat for a given species and assessing how much water, food, or space that particular species needs. Someone raising plants for food needs to recognize signs of disease, too. Though I emphasize the domestication of animals here, I don't deny the highly significant role played by the domestication of staple food crops in various ecosystems and human groups.

Domesticating animals provided a new sort of benefit. They were living tools first and meat sources later, only when their useful lives were over or circumstances required. The crucial importance of animal domestication in modern life shows that our relationsip with animals selected for a set of communication skills and abilities to observe, draw conclusions, and make connections among different observations that had been increasingly important since at least 2.6 million years ago. The relationship between such skills and the modern behaviors that characterize humanity is clear.

16

Riding into the Sunset

HORSES IN PARTICULAR provided an enormous advantage to those who domesticated them. Like the cowboys in many a movie—or the Comanche fighting against sedentary farmers—warriors from mounted cultures attacked their enemies swiftly and often won, galloping off into the sunset with booty or captives strapped to their saddles. Unmounted foes have little chance again horsemen.

Sandra Olsen argues that the horse made a greater impact on human society than any other animal. Before the domestication of horses, most people carried goods, cargo, and possessions on their backs. This strictly limits the mobility of a people and their options for trade. But once humans had horses, the situation was radically different. "Horses are swift of foot, can easily support one or two human passengers, carry heavy loads, and survive on very poor quality vegetation," Sandra says. "They were our first form of rapid transit."

The horse was enormously useful in transforming life in the Old World and, later, in the New World. There were unintended consequences however. Around 1300 AD, the presumed descendants of the Botai came riding back into Europe as the fearsome Mongol army headed by Genghis (or Chinggis) Khan and his descendants. Those consummate horsemen carried with them another spoil of war and another benefit of domestication: the plague. More clearly than almost any other example in history, the story of the Mongol hordes and the spread of the plague proves the deep biological importance of the intimacy between animals and humans.

Only people who had a way to transport masses of humanity and the goods they needed to survive could succeed at a military campaign like that carried out by the Mongol army. Quite simply, the Mongol Empire could not have been conquered or managed without domestic horses. And although killing their enemies through disease was hardly an intentional strategy of the Mongols at the outset, the rapid spread of a deadly zoonosis—a disease transmitted from animals to humans—would not have been possible without the intimate living arrangements of humans and various domestic animals. Disease carried by mounted horsemen proved a formidable warrior.

Genghis Khan was a Mongol horseman, born about 1160 AD into an obscure family. He welded the tribal horsemen of Asia into an almost unstoppable cavalry with a bloodlust for warfare. By 1206, he was the "Universal Ruler" or the Genghis Khan of the tribes of the Central Asian plateau (now northern China)— the Tartars, Naimans, Merkits, Uighurs, Mongols, and Keraits— uniting age-old enemies into the Mongol Empire through craft, cunning, and military brilliance. Each conquest of a kingdom or territory was followed by another. At the time of Genghis Khan's death in 1227, his twenty-one-year reign had expanded Mongol lands to include Tibet and northern China. According to Genghis's wishes, his sons each took over a khanate or administrative region and pursued the same mandate of rapid attack and conquest. By 1270, the Mongol Empire included southern China and Korea, the Persian lands of the conquered Sunni Muslim Khwarezm-Shah dynasty, much of Russia, Georgia, Poland, Hungary, the Balkans, and the Volga Bulgaria. At its height, the Mongol Empire was so large—stretching from the Yellow Sea of China to the Mediterranean of Europe—that no other continuous land empire in history has ever surpassed it. As far as the steppe-lands and prairies of Eurasia extended, the Mongol army went too, because of the outstanding horsemanship of its warriors.

One consequence of Mongol conquest was a rigorously enforced Pax Mongolica, which protected the key trading route known as the Silk Road and maritime trading routes connected to the Silk Road at importance junctures. Trade flourished along the newly safe conduit, carrying ideas like meritocracy, inventions like gunpowder, luxury goods like silk, and travelers like Marco Polo from one culture to the next. The first recorded outbreak of the plague, caused by the *Yersinia pestis* bacterium, was in China in 224 BC, long before the Pax Mongolica. At that time, settlements were relatively isolated and travelers few, so deadly outbreaks were confined by geography and eventually ran out of victims. But once the Silk Road was secured, the plague bacterium hitched a ride—unseen and probably unsuspected— with the caravans of goods and people. Where the Mongol army went, so went the plague.

According to a fourteenth-century writer, Gabriele de' Mussi, a critical event was the several years' siege of Caffa (also spelled Kaffa) in the Crimea, a key trading port on the Black Sea. After a year or two, the Tartar troops from the Mongol army camped outside the city gates began to suffer from plague. Normally, perhaps, the army had avoided the plague by being constantly on the move, so that no critical mass of sick humans and infected fleas could build up. But years of camping in one place took its toll on hygienic arrangements and forced the usually nomadic warriors to live like urban citizens. De' Mussi's vivid description brings the reality of the situation home. Thousands of soldiers were killed by the disease every day, "as though arrows were raining down from heaven to strike and crush the Tartars' arrogance," he wrote. "All medical advice and attention was useless; the Tartars died as soon as the signs of disease appeared on their bodies: swellings in the armpit or groin caused by coagulating humours, followed by a putrid fever."

The Tartars were ordered by their leaders to place the corpses in trebuchets and fling them over the city walls to kill the besieged within. No one could escape the stench of rotting corpses or the exposure to the deadly disease they carried. The few who escaped Caffa by boat still could not get away from the disease because they carried it with them. They spread the plague to Genoa, Venice, Marseilles, and other ports, infecting the inhabitants in each town. The disease reached Paris in 1348 and London in 1349. From there it spread inland, becoming one of the worst pandemics of all time, the Black Death.

Death tolls were enormous, accounting for perhaps 50 percent of the population of China, 20 percent of the populations of England and Germany, and 70–80 percent of the populations of Mediterranean Europe. Untreated plague, today, has a mortality rate of 60–100 percent. Though various scholars argue about whether the Black Death was the bubonic plague, the pneumonic plague (a form spread from person to person without an intervening vector), or a third as yet unknown form of hemorrhagic fever, the effect was the same. The disease spread rapidly, killed most individuals who caught it, and could not be effectively treated with the medical knowledge of the times.

So, what does this tale reveal about human history and evolution? This instance, so plainly recorded in contemporary documents, was directly dependent upon the close connection between animals and humans. Without horses for rapid transport of armies and traders, the plague would have erupted and died out locally. Horses were a convenient and very timely mechanism for spreading zoonoses, as airplanes are today. Without humans who lived in permanent, dense settlements in intimate association with domestic animals and unintended domesticates, as they might be called, like rats and mice, the spread of pestilence from rodent to flea to human would have been limited and relatively rare. But during the process of domestication, humans and the animals they domesticated became intertwined, both behaviorally and biologically.

The story of the Black Death is not an isolated instance. The second worst pandemic in historic times was another zoonosis: the Spanish flu that killed many more people in 1918–19 than had died in the entire First World War, perhaps 100 million people worldwide. This virulent strain of flu, identified scientifically as H1N1, is believed to have arisen among domestic fowl or pigs and spread to their owners and handlers. The domestic animals served as intermediate hosts for the virus. However, recent experiments with resurrected flu viruses from 1918 spread extremely rapidly and fatally among mice, indicating that they might have been intermediate hosts as well.

Other zoonoses have been nearly as virulent. Smallpox, measles, typhus, yellow fever, HIV, diphtheria, bubonic and pneumonic plague, the newly emerging SARS (severe acute respiratory syndrome), and avian and swine flu can all be securely traced to animal vectors. Because we live with domestic animals and continue to take in and nurture wild animals, humans are singularly vulnerable to the rapid spread of such diseases. Over half of the pathogens known to infect humans originated as zoonoses—the percentage is higher among those that pose the most serious health threats—and 60 percent of emerging infectious diseases are zoonotic.

Epidemiologists predict that the next round of severe, emerging diseases is likely to come from overcrowded areas of the tropical world where humans are being forced into closer contact with wild animals by overpopulation. More and more of the diseases from which humans suffer have their origin in the animals we live with. Even as humans have gained great advantages from living with animals and being intimately connected to animals, we have also suffered an increased vulnerability to particular types of diseases.

A review of human infectious diseases by Nathan D. Wolfe of the Johns Hopkins Bloomberg School of Public Health, Claire Panosian Dunavan of the University of California at Los Angeles Medical Center, and Jared Diamond of the University of California at Los Angeles considered twenty-five major human infectious diseases and their origins, insofar as those were known. They concluded that more diseases of temperate climates arose in the Old World than the New World because nearly all of the major domestic species originated in the Old World.

The llama and alpaca are the sole livestock species domesticated in the New World, and they have not infected humans with any pathogens. Why not? Traditionally, its geographic range was confined to the Andes, putting a limit (albeit large) on the territory through which any zoonosis could spread. Too, llamas and alpacas are handled more distantly than many other domestic animals; they are not milked and they are not kept inside or cuddled.

What this means is that long, intimate exposure between humans and an animal species is needed if an animal disease is going to jump the species barrier and spill over into humans. Onetime exposure to the pathogen is not enough.

Wolfe and his colleagues identify five evolutionary stages in the transition from an animals-only disease to a humans-only disease. Not all infectious agents evolve to become stage 5 diseases:

- Stage 1 diseases are microbes in animals that are not transmitted to humans under natural conditions, like most forms of malaria, which require an intermediate host (mosquitoes).

- Stage 2 is typically an animal pathogen that is transmitted from animals to humans (a primary infection) but not from human to human (a secondary infection). Rabies is a good example.

- Stage 3 pathogens are transmitted *to* humans but are only occasionally transmitted *between* humans, so that outbreaks of the disease die out unless people are reinfected from an animal host. Monkeypox virus is a stage 3 pathogen that moves from monkeys to humans, but not further.

- Stage 4 diseases exist in animals but also undergo long periods of human-to-human secondary infection, such as cholera, the scourge of refugee camps.

- Stage 5 is a pathogen that has evolved to be exclusive to humans, such as syphilis or smallpox.

The living tools that humans created out of a few of the animals around them have given us new opportunities, making it possible for humans to move into new habitats and new ways of life. Domesticating animals has been highly beneficial to our species and has in many cases increased the numbers and health of the animals we have domesticated. As Stephen Budiansky argues in *The Covenant of the Wild*, horses are but one example of a species that would almost certainly be extinct had they not been domesticated. So, although all humans do not behave kindly toward animals, we have co-evolved with the ones we have domesticated, and both species in the process have reaped benefits from the arrangement.

The price of the living tools we have created, however, is our greatly increased vulnerability to the diseases that affect (and infect) those tools. The animals with which we live, whose wool makes our cloth, whose hides make our leather, whose backs carry our goods and ourselves, also give us parasites, bacteria, and viruses. As we live with these animals, we have given the pathogens that co-exist with them a wonderful opportunity to evolve ways to jump the species barrier and live in us.

From the pathogen's point of view, the more hosts the better. But in order to attack a new species, the pathogen has to adapt its genetic structure to suit ours, or at least to evade our immune system. Since pathogens reproduce very rapidly, suitable adaptations can arise rapidly through random mutation, as long as the exposure to humans continues. And, though our immune systems can also evolve ways to defeat the invading pathogens, humans reproduce much more slowly than pathogens and thus evolve infinitely slower. All else being equal, the pathogens will win the adaptation race every time.

Today, as population densities continue to rise and humans are encroaching upon areas inhabited by wildlife, those animals, too, have begun to come into frequent enough contact with humans to pass pathogens to us. Where there are lots of humans living closely packed together, and where those human populations are still growing rapidly, humans are forced into new habitats and into new intimacy with animals even though those animals are not domestic or even tame. Which zoonoses successfully make the transition to infecting humans depends on a number of variables, including how many species serve as reservoirs for the pathogen, how many of the individuals in the reservoir species are infected, how good the vector—such as a mosquito—is at transmitting the disease, and how humans behave. For example, monkeypox is a virulent virus, but if no one ate or handled dead monkeys (a major type of bushmeat), then transmission would be very low. Another issue is how closely related the reservoir species are to humans. Species that are more closely related to humans—such as primates—are more likely to pass their zoonoses on to us simply because the genetic differences between them and us are fewer. Although primates make up only about 0.5 percent of all vertebrate species in the world, they have contributed 20 percent of the major human diseases, including AIDS, the Ebola virus, Marburg's disease, and herpes virus B.

Most new infectious diseases are zoonoses caused by pathogens from wildlife, not domestic animals. The density and growth of human populations are strong predictors of the location of emerging infectious diseases. Thus, by domesticating animals we have opened new opportunities for ourselves and for the animals we have domesticated, but also for the pathogens that infect them (and us).

17

The Animal Connection in the Modern World

WE ARE NO LONGER a world of farmers. Two hundred years ago, 90 percent of the population of the United States farmed; now only 2 percent do. Nonetheless, even in a heavily industrialized nation like the United States, our farming heritage is still strong. Aside from the almost mythical status of the good life on the farm that is popular in America, the actual statistics are impressive. In 2009, 40.8 percent of the total land acreage in the United States was farmland. Ninety-eight percent of all U.S. livestock farms are family-run; a similar percentage of family-run farms occurs in China.

We think of our society as industrialized, but do we live in a post-animal world? Have we become disconnected from animals and their importance in our lives as fewer people work in agriculture?

Hardly. A truly enormous number of people choose to live in daily association with animals, either in the livestock industry or as pets, now the animals that are most commonly intimately involved with humans. The statistics on pet ownership in diverse countries are clear evidence of the persistent importance of the human-animal link in the modern world. Put quite simply, humans have apparently evolved to need contact with animals.

Annual expenditures on pets are huge. For example, in 2007, Americans spent more than $41 billion on their pets, Australians expended $4.63 billion, Britons forked over $2.62 billion, Japanese pet owners parted with $10 billion, and Chinese laid out $995 million. Even people in dire circumstances—the homeless who have fled their homes because of natural disasters—make tremendous efforts to bring their pets with them.

Pets are not only expensive, they are numerous. There are 69 million pet-owning households (63 percent of all households) in the United States or Australia and 37 million (47 percent of all households) in the United Kingdom. In both the United States and the United Kingdom, the proportion of households with pets is larger than that with children. The number of dogs in Japan exceeds that of children under the age of twelve.

Why?

The answer most commonly given in surveys is that we have pets for companionship or because we love animals.

I would argue that we love animals and find companionship and joy in their presence because we have evolved to be connected to animals. Our connection with animals has been critical to our evolutionary survival for millions of years, and despite huge changes in our mode of subsistence, we still have a deep need to be involved with animals. The perceived intensity of this need varies from individual to individual, so I know some readers will scoff at this assertion. But a remarkable number of humans choose to live with animals, feel that their life is incomplete without them, and regard their animals as family members or loved ones. The need to be in contact with animals is still there in our species as a whole. Nothing else makes sense of the energy, money, emotion, and effort we spend on being with animals.

If being connected to animals is a genetically based behavior typical of humans, then there ought to be observable benefits to being with animals—and there are. Pets are not simply the leftovers of the domestication process kept around for some whimsical reason; they fulfill a need in our lives. Pets—"companion animals" in today's jargon—influence and improve human health not only in animal-loving, rich countries but cross-culturally. Many researchers have examined the interactions between pets and humans and explored their enormous value in therapeutic settings. Pets bring a unique set of benefits to their humans. Pet-owning individuals have better health, more social contacts, more exercise, and a better outlook on life than non-pet owners. Simply being with a pet lowers the heart rate, lowers cholesterol, and lowers anxiety in most people. Being with pets also raises oxytocin levels, the hormone that is key in bonding with our own infants and our mates. Increased oxytocin levels produce a feeling of calm and peacefulness, and heighten our sensitivity to nonverbal communication. Meg Daley Olmert, in her book *Made for Each Other*, suggests that oxytocin was the "main biological ingredient" underlying animal domestication and our affection for animals.

Pets and other domestic animals are also powerful agents for improvement for the physically and mentally impaired, and for those engaged in antisocial behaviors such as dropping out of school, crime, and drug taking. In *The Last Child in the Woods*, Richard Louv has proposed that the lack of exposure to nature in the lives of children in the industrialized world is directly linked to the rise in attention disorders, obsesity, behavioral disorders, anxiety, and depression. He has coined a phrase, nature-deficit disorder, that, though not an accepted clinical diagnosis, resonates with many people. Not surprisingly, even a slight increase in exposure to the natural world increases the resilience of humans to stress.

Working with animals is particularly effective in overcoming or lessening many physical and psychological difficulties because humans are genetically programmed to relate to animals and to communicate with them. Another way of stating this is that working with animals releases oxytocin and makes people calmer, more sociable, and more trusting. Working with animals helps poor readers develop the confidence to read aloud, increases the survival rate after heart attacks, improves social skills among autists, and diminishes the effects of ADHD (attention-deficit/hyper-activity disorder). But living with animals is no cure-all.

Many people cite the unconditional love that animals impart as the cause for the popularity of pets, and it is certainly an important factor. But if you look deeper, an expectation of unconditional love cannot have been the initial cause for domesticating animals. Nobody with the least knowledge of wolves would take in a baby wolf in hopes of receiving unconditional love. There is no guarantee that a tamed wolf or any other wild animal will refrain from outright attack, as many people who have foolishly bought exotic species as pets have found. The examples are myriad of people raising exotic animals, only to be seriously injured by the very animal on which they have lavished so much love. Even one of the lions raised by Joy and George Adamson (not Elsa, the lioness made famous by *Born Free*) attacked a child and then killed the Adamsons' cook, who had fed this very lion every day. These tragic outcomes are caused by a fundamental confusion between taming and domestication. A few years of kindly contact with humans cannot erase the predatory (or defensive) instincts bred into an animal for millennia.

Clearly, part of the basis of our intimacy with tame or domestic animals involves physical contact. People who work with animals touch them. It doesn't matter if you are a horse breeder, a farmer raising pigs, a pet owner, a zoo keeper, or a veterinarian, we touch them, stroke them, hug them. Many of us kiss our animals and many allow them to sleep with us. We touch animals because this is a crucial aspect of the nonverbal communication that we have evolved over millennia. We touch animals because it raises our oxytocin levels—and the animal's oxytocin levels. We touch animals because we and they enjoy it.

The same evolutionary tendency to be involved with animals also accounts for other immensely popular activities. Humans spend a lot of time and money visiting zoos, wildlife parks, and animal rehabilitation centers, hiking in wilderness areas, going on ecotourism trips to watch whales, wild macaws, or howler monkeys, canoeing up the Amazon, or watching the vast migrations on the Serengeti Plains of Africa. The temptation to touch or hold a wild animal is very strong. At a less adventurous level, a huge number of humans watch movies and television programs about animals such as *The March of the Penguins* or *Life of Mammals*. We buy vast amounts of posters, calendars, artwork, jewelry, and clothing with animal designs. Thousands of humans contribute to or work for conservation programs, animal rights groups, animal shelters, and the like. We put pictures of animals on our desks, our walls, our screensavers, and in our hearts. Taken as a whole, the widespread human interest in animals shows that we have still have a compelling need to connect with them.

Time spent with animals heightens our senses, makes us more aware of color, shape, sound, smell, and movement. Animals act predictably, or unpredictably, renewing our sense of surprise and wonder. They calm our hearts and our souls.

And what of people who do live without animals, perhaps in sterile, high-rise apartment buildings in cities where there is little trace of green, much less of wildlife? I think it could be argued that the very qualities the animal connection has selected for in humans, over millions of years, are diminished by such environments. We lose compassion, empathy, and communication skills; people live isolated lives, with fewer close emotional ties or supportive human networks; we become more robotic, less caring, and less spiritual. Anxiety heightens, alienation increases, tension builds, and there is little release or healing. Children grow up without spending time outside simply exploring, watching, playing, or daydreaming. They are reared in confined safe spaces; they are entertained, rather than learning to follow their curiosity and entertain themselves; sterile, germ-free worlds dull their senses and stifle their imaginations. Soon they require ever-increasing, ever more extreme stimulation to provoke their attention.

The post-animal world, if we choose to live in it, is a fearsome place that threatens to destroy the very best qualities of humankind.

AT THE BEGINNING of this book, I presented a challenge. I identified my hypothesis: that our connection to animals is ancient and fundamentally important because it drove humans to our three great behavioral advances: the making of stone tools; the origin of language and symbolism; and the domestication of animals. I have traced the story of human evolution from 2.6 million years ago to modern times, presenting evidence and making observations that I believe show compellingly that our connection with animals was instrumental in shaping the human organism we are today. From the first stone tool to the origin of language and the most recent living tools, our involvement with animals has directed our course.

I am certainly not the only person to notice the importance of animals' influence on humans. For one, Jared Diamond has called "plant and animal domestication the most important development in the past 13,000 years." In *The Covenant of the Wild*, Stephen Budiansky has controversially argued that the animal species that became domesticated chose domestication— that the change from wild to domesticated was a sort of mutual covenant between humans and animals. Temple Grandin and Catherine Johnson have written movingly on the theme that *Animals Make Us Human* and how her connection to animals helped Grandin—an autist—cope with life. Barbara J. King has explored the deep spiritual connection between animals and people in *Being with Animals*, and Hal Herzog has pon-

dered the puzzling and often contradictory ethics of our relationships with animals in *Some We Love, Some We Hate, and
Some We Eat*. Richard Louv mourns the deadening effects of
being deprived of exposure to nature—bugs, trees, animals,
birds, slime, and dirt—on our children. Within this community
of thinkers, scholars, and writers, I am perhaps the only one
to have traced the complex relationship between animals and
humans in such detail over millions of years, limning a trajectory that explains why our interactions with animals continue
to affect us so deeply.

I have argued here that the animal connection has been one
of the key influences on the human species as we have evolved.
Our increasingly intimate interactions with animals have
shaped our lifestyle and driven us to become more observing,
more empathetic, better at communicating, more tolerant, and
more able to compromise or negotiate. Simply paying attention to other animals—focusing on them, watching what they
do, how they live, and what they need, gathering that intensely
important information and communicating it to others—has
had an incredibly large effect on our evolution. If our connection with animals has made us human, what will it do for us in
the future?

Sadly, I see little evidence that urban planning or architec-
ture makes room for interactions with animals. Populations are
too large. Dwellings are often packed in at maximum density
and planned green spaces tend to be too small to support any
sort of wildlife except perhaps a few roaches, squirrels, and
birds well adapted to urban life. Many children already grow up
with no firsthand knowledge of animals, no sense of where meat
comes from except the grocery store, and little understanding

of animal (and hence human) behavior. These are the ones suffering from Louv's nature-deficit disorder. Yes, you can house more people more economically in high-rise, concrete structures, especially if you expend no dollars on beauty, form, or shape. But many city apartments are poorly suited to most pets (I would argue, to most humans) and living in them deprives people of the physical, emotional, and mental comforts the animal connection provides, points raised so effectively by Jeffrey Moussaief Masson in *The Dog Who Couldn't Stop Loving*. Our need to live with animals, touch them, or work with them is embedded in our genes, if my hypothesis is correct. We cannot cast off our need for animals or our pleasure in them lightly or voluntarily and remain the same.

One example of this is so simple and so clear that you may need no other. As large parts of the world become more urban, family ties and lifelong personal relationships weaken. Many elderly people are no longer cared for by family members but by strangers, in managed care facilities or nursing homes. But one of the most obvious sources of difficulty with this arrangement has been largely overlooked. A recent study in the United Kingdom concluded that, as elderly or disabled people are removed from their homes and moved into care facilities, a major cause of their stress, unhappiness, and physical and mental deterioration is being parted from pets. Of course, it is awkward to allow pets in such facilities. They must be fed, cared for, cleaned up after, and exercised—just like the human inhabitants. Such inconvenience is outweighed by the better health, better attitudes, diminished incidence of depression and anxiety, and increased sociability among the humans. The evidence was so overwhelmingly clear that the British House of Commons passed a bill in

2010 to increase the number of residential care facilities that allow the elderly to bring their pets with them.

The findings of that study only reinforce my conviction that the importance of animals in our lives has been grossly underestimated in the modern world. And yet, how thoroughly we humans are captivated on the rare occasion when the natural world bursts into the consciousness of the urban dweller. The lovely story of Pale Male, a red-tailed hawk who roosted, courted, and raised his young on an apartment building in New York City, touched millions of hearts. How readily we become enchanted with baby pandas at the zoo, how effective are photos of oil-drenched pelicans and sea otters being rescued at stirring our emotions and increasing legislative pressure for more regulation of the oil-drilling industry.

Despite the difficulties in finding pet-friendly accommodations, many people who live in big cities own pets "for companionship," a need apparently not fulfilled by humans. The purr of a cat, the wagging tail of a dog may be the warmest and most fulfilling interaction of a person's entire day. Evidence like that, common sense, and the long trajectory of human-animal interactions presented here would suggest we must take great care to preserve our ability to live with animals.

I believe that interacting with animals on an intimate basis in large part caused us to evolve sophisticated tools and enhanced communication skills, including the invention of language itself. It was animals who taught us that others—even other species—have emotions, needs, and "thoughts." It was animals who selected for the vital skills of empathy, understanding, and compromise in the human lineage. Developing those skills and our longing for an intimate connection with animals is a profoundly important aspect of human behavior and human biology.

If we phase animals out of our world—if we disregard their importance and relevance to our lives—we will relax the selective advantage that the animal connection has conferred on communication and empathy with others.

Our world is becoming global. Many of us are in daily contact with people from other parts of the world, other cultures, other language groups. As globalization proceeds, we will need, urgently, every ounce of empathy, tolerance, and communication skill we can muster. This is what our connection with animals has given us; this is what we need so badly for the future.

Acknowledgments

Writers do not live by words alone, at least not this one.

For support, suggestions, editorial comments, sympathy, and intellectual stimulation, I thank first and foremost my husband, Alan Walker, and then many friends and colleagues: Zeray Alemseged, Lucinda Backwell, Bob Brain, Nancy Marie Brown, George Chaplin, Melissa Deines and her zoo, Francesco d'Errico, Bill Fields, Mietje Germonpré, Cheryl Glenn, Haskel Greenfield, Hal Herzog, Brian Hesse, Leslea Hlusko, Peter Hudson, Nina Jablonski, Jennie Jin, Barbara Kennedy, Barbara J. King, Curtis Marean, Gigi Marino, Jeffrey Moussaieff Masson, Lee Newsom, Sandra Olsen, Jon Olson, Kathy Schick, Sue Savage-Rumbaugh, Sileshi Semaw, Mary Stapleton, Nick Toth, Chris Wadman, Simon, Shellene, Bryn, and Meghan Walker, Bob Wayne, David Webster, Tim White, as well as Kanzi, Panbanisha, Jennifer and Zelda Shipman. Angela von der Lippe gave a helpful and sympathetic reading to the manuscript and her assistant, Laura Romain, was wonderfully efficient. My late agent, Ralph Vicinanzi, managed to sell this book in a time when nobody was buying books and Michelle Tessler, my new agent, stepped in gracefully after Ralph's death. Thanks!

Notes

Prologue

15 **given by the King to the White Rabbit**—L. Carroll, *Alice's Adventures in Wonderland* (1865; London: Macmillan & Co., 1913), p. 187.

Chapter 1

20 **Orrorin tugenensis**—Senut, B., et al. 2001. "First hominid from the Miocene (Lukeino Formation, Kenya)," *Comptes Rendus de l'Académie des Sciences. Series IIA—Earth and Planetary Science*, 332 (2): 137–44.

20 **Sahelanthropus tchadensis**—Brunet, M., et al. 2002. "A new hominid from the Upper Miocene of Chad, Central Africa," *Nature*, 418:145–51.

20 **Ardipithecus kadabba**—Haile-Selassie, Y., Suwa, G., and White, T. D. 2004. "Late Miocene teeth from Middle Awash, Ethiopia, and early hominid dental evolution," *Science*, 303:1503–05.

21 **Ardipithecus ramidus**—White, T., et al. 2009. "*Ardipithecus ramidus* and the paleobiology of early hominids," *Science*, 326:75–86; Ashfaw, B., White, T. D., and Suwa G. 1994. "*Aus-*

tralopithecus ramidus, a new species of early hominid from Aramis, Ethiopia," *Nature,* 37:306.

22 **what modern cheetahs leave behind**—C. K. Brain, *The Hunters or the Hunted? An Introduction to African Cave Taphonomy* (Chicago: University of Chicago Press, 1983).

26 **the microscopic wear on their teeth**—Ungar, P. S., Grine, F., and Teaford, M. F. 2008. "Dental microwear and diet of the Plio-Pleistocene hominin *Paranthropus boisei,*" *PLoS ONE,* 3 (4): e2044.

28 **cutmarks made by stone tools**—McPherron, S., et al. 2010. "Evidence for stone-tool-assisted consumption of animal tissues before 3.39 million years ago at Dikika, Ethiopia," *Nature,* 466:857–60.

29 ***Homo habilis***—Leakey, L. S. B., Tobias, P. V. T., and Napier, J. R. 1964. "A new species of the genus *Homo* from Olduvai Gorge," *Nature,* 202:7–9.

Chapter 2

34 **obsidian blades are sharper**—Buck, B.A. 1982. "Ancient technology in contemporary surgery," *Western Journal of Medicine,* 136 (3): 265–69; Disa, J. J., Vossoughi, J., and Goldberg, N. H. 1993. "A comparison of obsidian and surgical steel scalpel wound healing in rats," *Plastic and Reconstructive Surgery,* 92 (5): 884–87.

40 **the oldest known archeological sites**—Semaw, S. 2000. "The world's oldest stone artefacts from Gona, Ethiopia: Their implications for understanding stone technology and patterns of human evolution between 2.6–2.5 million years ago," *Journal of Archaeological Science,* 27:1197–1214; Stout, D., et al. 2010. "Technological variation in the earliest Oldowan from Gona, Afar, Ethiopia," *Journal of Human Evolution,* 58:474–91.

41 **technological stasis**—S. Semaw, personal communication to the author.

41 **Specimens from Gona and at sites at Bouri**—de Heinzelin, J., et al. 1999. "Environment and behavior of 2.5-million-year-old Bouri hominids," *Science*, 284:625–29; Dominguez-Rodrigo, M., et al. 2005. "Cutmarked bones from Pliocene archaeological sites at Gona, Afar, Ethiopia: Implications for the function of the world's oldest stone tools," *Journal of Human Evolution*, 48 (2): 109–21.

42 **"Cutmarks!"**—Potts, R., and Shipman, P. 1981. "Cutmarks made by stone tools on bones from Olduvai Gorge, Tanzania," *Nature*, 291: 577–80; Shipman, P., and Rose, J. J. 1983. "Early hominid hunting, butchering and carcass-processing behaviors: Approaches to the fossil record," *Journal of Anthropological Archaeology*, 2:57–98; Shipman, P., 1986. "Studies of hominid-faunal interactions at Olduvai Gorge," *Journal of Human Evolution*, 15:691–706; and Olsen, S. L., and Shipman, P. 1988. "Surface modification on bone: Trampling versus butchery," *Journal of Archaeological Science*, 15:535–49.

44 **In subsequent years**—See Blumenschine, R., and Selvaggio, M. 1988. "Percussion marks on bone surfaces as a new diagnostic of hominid behavior," *Nature*, 333:763–65; Blumenschine, R. 1995. "Percussion marks, tooth marks, and experimental determinations of the timing of hominid and carnivore access to long bones at FLK Zinjanthropus, Olduvai Gorge, Tanzania," *Journal of Human Evolution*, 29:21–51; Blumenschine, R., Marean, C., and Capaldo, S. 1996. "Blind tests of inter-analyst correspondence and accuracy in the identification of cut marks, percussion marks, and carnivore tooth marks on bone surfaces," *Journal of Archaeological Science*, 23:493–507; and Pickering, T. R., and Egeland, C. P. 2006. "Experimental patterns of hammerstone percussion damage on bones: Implications for inferences of carcass processing by humans," *Journal of Archaeological Science*, 33:459–69.

45 **crocodiles, studied by Jackson Njau**—Njau, J. K., 2006. *The Relevance of Crocodiles to Oldowan Hominin Paleoecology at*

Olduvai Gorge, Tanzania. Dissertation submitted to the Graduate School, New Bruswick, of Rutgers, the State University of New Jersey.

46 **early tools were used on plant materials**—Keeley, L., and Toth, N. 1981. "Microwear polish on early stone tools from Koobi Fora," *Nature*, 293:464; Dominguez- Rodrigo, M., et al. 2001. "Woodworking activities by early humans: A plant residue analysis on Acheulian stone tools from Peninj (Tanzania)," *Journal of Human Evolution*, 40 (4): 289–99.

47 **More flakes equals more cutmarks**—Shipman, P., 1986. "Studies of hominid-faunal interactions at Olduvai Gorge," *Journal of Human Evolution*, 15:691–706.

48 **Prey size at Olduvai**—Ibid. See also R. Potts, *Early Hominid Activities at Olduvai* (New York: Aldine de Gruyter, 1988); Fernandez-Jalvo, Y., et al. 1998. "Taphonomy and palaeoecology of Olduvai Bed I (Pleistocene, Tanzania)," *Journal of Human Evolution*, 34:137–72; and Fernandez-Jalvo, Y., Andrews, P., and Denys C. 1999. "Cut marks on small mammals at Olduvai Gorge Bed I," *Journal of Human Evolution*, 36:587–89.

48 **Carnivores focus their hunting**—Owen-Smith, N., and Mills, M. G. L. 2008. "Predator–prey size relationships in an African large-mammal food web," *Journal of African Ecology*, 77:173–83; Hertler, C., and Volker, R. 2008. "Assessing prey competition in fossil carnivore communities—A scenario for prey competition and its evolutionary consequences for tigers in Pleistocene Java," *Palaeogeography, Palaeoclimatology, Palaeoecology*, 257:67–80; H. Hemmer. "Notes on the Ecological Role of European Cats (*Mammalia: Felidae*) of the Last Two Million Years," in E. Baequedano and S. Rusto Jarra, eds. *Miscelánea en homenaje a Emiliano Aguirre.* Vol. II: *Paleontología* (Alcalá de Henares: Museo Arqueológico Regional, 2004); C. Nwokeji, personal communication to the author.

50 **Paranthropines, australopithecines, and the earliest humans**—McHenry, H., and Coffing, K. 2000. "*Australopithecus*

to *Homo;* Transformations in body and mind," *Annual Review of Anthropology*, 29:125–46.

52 **The preferred prey for any carnivore**—Data from Bunn, H., and Kroll, E. M. 1986. "Systematic butchery by Plio-Pleistocene hominids at Olduvai Gorge, Tanzania," *Current Anthropology*, 27:431–52.

54 **Chimps are absolutely hopeless carnivores**—Stanford, C. B., et al. 1994. "Patterns of predation by chimpanzees on red colobus monkeys in Gombe National Park, Tanzania, 1982–1991," *American Journal of Physical Anthropology*, 94:213–28; C. B. Stanford, *Chimpanzee and Red Colobus: The Ecology of Predator and Prey* (Cambridge, MA: Harvard University Press, 1998), and Stanford, C. B. *The Hunting Apes: Meat-eating and the Origins of Human Behavior* (Princeton: Princeton University Press, 1999); Stanford, C. B. "The predatory behavior and ecology of wild chimpanzees" at www-rcf.usc.edu/~stanford/chimphunt.html, accessed June 14, 2010; Takahata, Y., Hasegawa, T., and Nishida, T. 1984. "Chimpanzee predation in the Mahale Mountains from August 1979 to May 1982," *International Journal of Primatology*, 5:213–33.

55 **Though olive baboons . . . their focal prey**—Hausfater, G. 1976. "Predatory behavior of yellow baboons," *Behaviour*, 56:44–68; S. C. Strum, "Processes and Products of Change: Baboon predatory behavior at Gilgil, Kenya," in R. S. O. Harding and G. Teleki, eds. *Omnivorous Primates* (New York: Columbia University Press, 1981), pp. 255–302; S. C. Strum, and W. Mitchell, "Baboon Models and Muddles," in W. G. Kinzey, ed., *The Evolution of Human Behavior: Primate Models* (Albany, NY: State University of New York Press, 1987).

55 **"Chimpanzees are limited"**—Plummer, T., and Stanford, C. 2000. "Analysis of a bone assemblage made by chimpanzees at Gombe National Park, Tanzania," *Journal of Human Evolution*, 39:345–65, p. 357.

Chapter 3

57 **a large proportion of cutmarks**—Bunn, H., and Kroll, E. M. 1986. "Systematic butchery by Plio-Pleistocene hominids at Olduvai Gorge, Tanzania," *Current Anthropology*, 27:431–52.

60 **Expensive Tissue hypothesis**—Aiello, L., and Wheeler, P. 1995. "The expensive tissue hypothesis," *Current Anthropology*, 36:199–211; Aiello, L., and Wells, J. C. K. 2002. "Energetics and the evolution of the Genus *Homo*," *Annual Review of Anthropology*, 32:323–38.

64 **There are only three evolutionary options**—Shipman, P., and Walker, A. 1989. "The costs of becoming a predator," *Journal of Human Evolution*, 18:373.

67 **a study of coyotes and wolves**—Berger, K. M., and Gese, E. M. 2007. "Does interference competition with wolves limit the distribution and abundance of coyotes?" *Journal of Animal Ecology*, 76:1075–85.

68 **there were eleven large carnivores**—B. Van Valkenburgh. "The Dog-Eat-Dog World of Carnivores: A Review of Past and Present Carnivore Community Dynamics," in C. B. Stanford and H. T. Bunn, eds., *Meat-eating and Human Evolution* (Oxford: Oxford University Press, 2001), pp. 101–21.

Chapter 4

73 **Osteodontokeratic culture**—Dart, R. 1957. "The osteodontokeratic culture of *Australopithecus prometheus*," *Transvaal Museum Memoirs*, no. 10; Dart, R. A. 1949. "The predatory implemental technique of *Australopithecus*," *American Journal of Physical Anthropology*, 7:1–38.

73 **Mary Leakey . . . in her definitive book**—See M. D. Leakey, *Olduvai Gorge*, Vol. 3: *Excavations in Beds I and II, 1960–1963* (Cambridge: University Press, 1971).

74 **many of the bone tools she had identified**—P. Shipman, and J. Rose, "Bone Tools: An Experimental Approach," in S. Olsen,

ed., *SEM in Archaeology*, BAR International Series 452 (Oxford: British Archaeological Reports, 1988), pp. 303–36; P. Shipman, "Altered Bones from Olduvai Gorge, Tanzania: Techniques, Problems, and Implications of Their Recognition," in R. Bonnichsen and M. Sorg, eds., *Bone Modifications* (Orono, ME: Center for the Study of Early Man, 1989), pp. 317–34.

77 **Four fascinating specimens**—Backwell, L., and d'Errico, F. 2004. "The first use of bone tools: a reappraisal of the evidence from Olduvai Gorge, Tanzania," *Palaeontologia Africana*, 40:95–158.

77 **But photos of crocodile damage**—Njau, J. and Blumenschine, R. 2006. "A diagnosis of crocodile feeding traces on larger mammal bone, with fossil examples from the Plio-Pleistocene, Olduvai Basin, Tanzania," *Journal of Human Evolution*, 50:142–62.

79 **Bob's specimens looked nothing at all like**—C. K. Brain and P. Shipman, 1993. "Bone Tools from Swartkrans," in C. K. Brain, ed., *Swartkrans: A Cave's Chronicle of Early Man* (Pretoria: Transvaal Museum Monograph No. 8, 1993), pp. 195–215.

81 **Lucinda and Francesco found a pattern**—Backwell, L., and d'Errico, F. 2001. "Evidence of termite foraging by Swartkrans early hominids," *Proceedings of the National Academy of Sciences, USA*, 98 (4): 1358–63.

83 **but so would eating lots of termites**—Sponheimer, M., et al. 2005. "Hominins, sedges, and termites: New carbon isotope data from the Sterkfontein Valley and Kruger National Park," *Journal of Human Evolution*, 48 (3): 301–12; d'Errico, F., Backwell, L., and Berger, L. 2001. "Bone tool use in termite foraging by early hominids and its impact on our understanding of early hominid behavior," *South African Journal of Science*, 97:71–75.

83 **"One hundred grams of rump steak"**—L. Backwell, personal communication to the author.

83 **40,000 termites a night**—Taylor, W. A., Lindsey, P. A., and Skinner, J. D. 2002. "The feeding ecology of the aardvark *Orycteropus afer*," *Journal of Arid Environments*, 50:135–52; Ohiagu, C. E. 1979. "Nest and soil populations of *Trinervitermes* spp. with particular reference to *T. geminatus* (Wasmann), (Isoptera), in

southern Guinea savanna near Mokwa, Nigeria," *Oecologia*, 40 (2): 167–78.

84 **possible bone tools from . . . Drimolen**—d'Errico, F., and Backwell, L. 2009. "Assessing the function of early hominin bone tools," *Journal of Archaeological Science*, 36:1764–73.

Chapter 5

87 **"Man is a social animal"**—Oakley, K. 1949. "Man the Tool-maker," *Bulletin of the British Museum*, p. 1.

88 **groundbreaking paper in *Nature***—Goodall, J. 1964. "Tool-using and aimed throwing in a community of free-living chimpanzees," *Nature*, 201:1264.

88 **"Now we must redefine"**—Leakey, L. S. B. 1964. Press conference in Washington, D.C., April 4, 1964.

89 **spears for hunting bushbabies**—Pruetz, J., and Bertolani, P. 2007. "Savanna chimpanzees, *Pan troglodytes verus*, hunt with tools," *Current Biology*, 17:412–17.

89 **Fifty-six different behaviors that might be considered**—Whiten, A., et al. 1999. "Cultures in chimpanzees," *Nature*, 399:682–85.

89 **a group of researchers led by Julio Mercader**—Mercader, J., et al. 2007. "4,300-year-old chimpanzee sites and the origins of percussive stone technology," *Proceedings of the National Academy of Sciences, USA*, 104 (9): 3003–48.

91 **"the possibility that humans could be"**—Ibid., p. 3045.

92 **"the stage of an unintentional production of debris"**—Delagnes, A., and Roche, H. 2005. "Late Pliocene hominid knapping skills: The case of Lokalalei 2C, West Turkana, Kenya," *Journal of Human Evolution*, 48:435–72, p. 436.

92 **"does *not* mimic"**—N. Toth, and K. Schick, "An Overview of the Oldowan Industrial Complex: The Sites and the Nature of Their Evidence. Introduction," in N. Toth and K. Schick, eds.,

The Oldowan: Case Studies into the Earliest Stone Age (Gosport, IN: Stone Age Institute Publications, 2007), p. 24.

93 **"The moment when a hominid"**—de Beaune, S. 2004. "The invention of technology: Prehistory and cognition," *Current Anthropology*, 45 (2): 139–62, p. 4.

94 **Chimps do not make tools**—For a review of the toolmaking and -using capacities of chimpanzees, see Whiten, et al., "Cultures in chimpanzees," pp. 682–85.

95 **observed chimpanzees in Fongoli**—Pruetz and Bertolani, "Savanna chimpanzees, *Pan troglodytes verus*, hunt with tools," 412–17.

Chapter 6

98 **She didn't say so**—S. Savage-Rumbaugh and W. Fields, "Rules and Tools: Beyond Anthropomorphism," in Toth and Schick, eds., *The Oldowan: Case Studies into the Earliest Stone Age*, p. 226.

98 **Both Nick and Kathy are expert**—The following account is taken from interviews with Toth, Schick, and Savage-Rumbaugh, and information in S. Savage-Rumbaugh and R. Lewin, *Kanzi: The Ape at the Brink of the Human Mind* (New York: John Wiley & Sons, 1994); Toth, N., et al. 1993. "*Pan* the tool-maker: Investigations into the stone tool-making and tool-using capabilities of a bonobo (*Pan paniscus*)," *Journal of Archaeological Science*, 20 (1): 81–91; and Schick, K., et al. 1995. "Continuing investigations into the tool-making and tool-using capabilities of a bonobo (*Pan paniscus*)," *Journal of Archaeological Science*, 26 (7): 821–32.

99 **was sensitive to failure**—Savage-Rumbaugh and Fields, "Rules and Tools: Beyond Anthropomorphism," in Toth and Schick, eds., *The Oldowan: Case Studies into the Earliest Stone Age*, p. 232.

100 **"an important outside female visitor"**—Ibid., p. 233.

103 **"Tool use in wild bonobos"**—de Waal, F. 1995. "Bonobo sex and society," *Scientific American*, 272 (3): 82–88, p. 83.

105 **Nick and Kathy collaborated with Sileshi Semaw**—See Toth, Schick, and Semaw, "A comparative study of the stone tool-making skills of *Pan, Australopithecus,* and *Homo sapiens,*" in Toth and Schick, eds., *The Oldowan: Case Studies into the Earliest Stone Age,* pp. 155–222.

105 **they are about five times stronger**—Walker, A. 2009. "The strength of great apes and the speed of humans," *Current Anthropology,* 50 (2): 229–34.

106 **Nick thinks the toolmaking bonobos**—N. Toth, personal communication to the author.

108 **"There seems to have been much more knapping"**—Davidson, I., and McGrew, W. 2005. "Stone tools and the uniqueness of human culture," *Journal of the Royal Anthropological Institute,* n.s., 11: 793–817, pp. 810–11.

Chapter 7

111 **An Acheulian site known as Gesher Benot Ya'akov**—Information on Gesher Benot Ya'akov is taken from: Alperson-Afil, N., et al. 2010. "Spatial organization of hominin activities at Gesher Benot Ya'aqov, Israel," *Science,* 326:1677–80; Goren-Inbar, N., et al. 2004. "Evidence of hominin control of fire at Gesher Benot Ya'aqov, Israel," *Science,* 304:725–27; and Goren-Inbar, N., et al. 2002. "Nuts, nut cracking, and pitted stones at Gesher Benot Ya'aqov, Israel," *Proceedings of the National Academy of Sciences, USA,* 99:2455–60.

116 **"formalized conceptualization of living space"**—Alperson-Afil, et al. 2010. "Spatial organization of hominin activities at Gesher Benot Ya'aqov, Israel," p. 1680.

Chapter 8

121 **"If an animal had a capacity"**—N. Chomsky, "Clever Kanzi," *Discover* (March 1991), p. 20.

123 **"the language faculty evolved in the human lineage"**—
Pinker, S., and Jackendoff, R. 2005. "The faculty of language:
What's special about it?" *Cognition*, 95:201–36, p. 204.

123 **define language as**—Noble, W., and Davidson, I. 1991. "The
evolutionary emergence of modern human behaviour: Language
and its archaeology." *Man*, n.s., 26 (2): 223–53, p. 224.

123 **which Chomsky emphasizes over communication**—N.
Chomsky, *On Nature and Language* (New York: Cambridge University Press, 2000), p. 75.

123 **Case studies concerning individuals**—Bates, E., and Goodman, J. C. 1997. "On the inseparability of grammar and the lexicon: Evidence from acquisition, aphasia and real-time processing,"
Language and Cognitive Processes, 12 (5/6): 507–84, p. 528.

124 **The textbook example is a girl named Genie**—Information
on Genie is taken from: Fromkin, V., et al. 1974. "The development of language in Genie: A case of language acquisition beyond
the 'Critical Period,'" *Brain and Language*, 1:81–107; Curtiss, S.
1979. "Genie: Language and cognition," *UCLA Working Papers
in Linguistics*, 1: 15–62; S. Curtiss, *Genie: A Psycholinguistic
Study of a Modern-Day "Wild Child"* (New York: Academic Press,
1977); R. Rymer, *Genie: A Scientific Tragedy* (New York: Harper,
1993); Jones, P. E. 1995. "Contradictions and unanswered questions in the Genie case: A fresh look at the linguistic evidence,"
Language and Communication, 15 (3): 261–80; and *Genie*, a
BBC Horizon documentary, 1994.

124 **"Mike paint" . . . "Teacher said"**—Curtiss, *Genie: A Psycholinguistic Study of a Modern-Day "Wild Child,"* pp. 35, 58.

126 **"Why does the paste"**—Fromkin, et al., "The development of
language in Genie: A case of language acquisition beyond the
'Critical Period,'" p. 128.

126 **He claims that the original function**—See T. Ingold, *The
Appropriation of Nature* (Manchester: Manchester University
Press, 1986), p. 70.

126 **Robin Dunbar . . . has hypothesized**—R. Dunbar, *Grooming,*

Gossip, and the Evolution of Language (Cambridge, MA: Harvard University Press, 1998).

127 **"A human language is a system"**—N. Chomsky, *Reflection of Language* (New York: Pantheon Books, 1975), p. 4.

129 **"Tissue paper blue rub face"**—Jones, P. E. 1995. "Contradictions and unanswered questions in the Genie case: A fresh look at the linguistic evidence." *Language and Communication*, 15 (3): 261–80, p. 269.

130 **"Aena to macha churen"**—D. Bickerton, *Language and Species* (Chicago: University of Chicago Press, 1990), p. 120.

131 **The work carried out . . . by Elizabeth Bates . . . and Judith C. Goodman**—Bates, E., and Goodman, J. C. 1997. "On the inseparability of grammar and the lexicon; Evidence from acquisition, aphasia and real-time processing," *Language and Cognitive Processes*, 12 (5/6): 507–84.

132 **In a second study**—E. Bates and J. C. Goodman, "On the Emergence of Grammar from the Lexicon," in B. MacWhinney, ed. *The Emergence of Language* (Mahwah, NH: Lawrence J. Erlbaum Associates, 1999).

134 **the call given by vervet monkeys**—D. Cheney and R. Seyfarth, *How Monkeys See the World* (Chicago: University of Chicago Press, 1990).

135 **"It's not that animals are too dumb"**—D. Bickerton, *Adam's Tongue: How Humans Made Language, How Language Made Humans* (New York: Farrar, Straus & Giroux, 2009), pp. 45–46.

136 **use the term *informavore***—J. Tooby, and I. DeVore, "The Reconstruction of Hominid Behavioral Evolution Through Strategic Modeling," in Kinzey, ed. *The Evolution of Human Behavior: Primate Models*, pp. 183–237.

Chapter 9

138 **"An approach is needed that"**—Noble and Davidson, "The evolutionary emergence of modern human behaviour: Language and its archaeology," p. 224.

138 **a sort of checklist of abilities**—C. Henshilwood, "Fully Symbolic *sapiens* Behaviour: Innovation in the Middle Stone Age at Blombos Cave, South Africa," in K. Boyle, D. Bar-Yosef, and C. Stringer, eds., *Rethinking the Human Revolution: New Behavioural and Biological Perspectives on the Origins and Dispersal of Modern Humans* (Cambridge: McDonald Institute for Archaeological Research, 2007), pp. 123–32; Bar-Yosef, O. 1998. "On the nature of transitions: The Middle to Upper Palaeolithic and the Neolithic Revolution," *Cambridge Archaeological Journal*, 8 (2): 141–63; Bar-Yosef, O. 2002. "The Upper Paleolithic Revolution," *Annual Review of Anthropology*, 31:363–93; Klein, R. G. 2000. "Archeology and the evolution of human behavior," *Evolutionary Anthropology*, 9:17–36; P. Mellars, "Technological Changes at the Middle-Upper Palaeolithic Transition: Economic, Social and Cognitive Perspectives," in P. Mellars and C. Stringer, eds. *The Human Revolution: Behavioural and Biological Perspectives on the Origins of Modern Humans* (Edinburgh: Edinburgh University Press, 1989), pp. 338–65; G. A. Clark, and J. M. Lindly, "The Case of Continuity: Observations on the Biocultural Transition in Europe and Western Asia," in ibid., pp. 626–76; White, R. K. 1982. "Rethinking the Middle/Upper Paleolithic transition," *Current Anthropology*, 23 (2): 169–92; Clark, G. A., and Lindly, J. M. 1991. "On paradigmatic biases and Paleolithic research traditions," *Current Anthropology*, 32: 577–87; and Hayden, B. 1993. "The cultural capacity of Neandertals: A review and reevaluation," *Journal of Human Evolution*, 24:113–46.

139 **The oldest recognizably modern skulls**—McDougall, I., Brown, F. H., and Fleagle, J. G. 2005. "Stratigraphic placement and age of modern humans from Kibish, Ethiopia," *Nature*, 433:733–36.

140 **Sally McBrearty of the University of Connecticut**—S. McBrearty, "Down with the Revolution," in Mellars, Boyle, Bar-Yosef, and Stringer, eds. *Rethinking the Human Revolution*, pp. 1343–51; McBrearty, S., and Brooks, A. 2000. "The revolution

that wasn't; A new interpretation of the origin of modern human behavior," *Journal of Human Evolution*, 39:453–563.

140 **"the key criterion for modern human behavior"**—Henshilwood, C. S., and Marean, C. W. 2003. "The origin of modern human behavior: Critique of the models and their test implications," *Current Anthropology*, 44 (5): 627–51.

140 **a *Homo habilis* skull from Sterkfontein**—Pickering, T. R., White, T. D., and Toth, N. 2000. "Brief communication: Cutmarks on a Plio-Pleistocene hominid from Sterkfontein, South Africa," *American Journal of Physical Anthropology*, 111 (4): 579–84.

141 **600,000 years ago at Bodo, Ethiopia**—White, T. D. 1986. "Cutmarks on the Bodo cranium: A case of prehistoric defleshing," *American Journal of Physical Anthropology*, 69:503–09.

141 **All of the Herto Bouri skulls bear cutmarks**—De Heinzelein, J., et al. 1999. "Environment and behavior of 2.5-million-year-old Bouri hominids," *Science*, 284:625–29.

141 **unmistakable burials of anatomically modern humans**—B. Vandermeersch, *Les Hommes Fossiles de Qafzeh (Israel)* (Paris: CNRS, 1981); Bar-Yosef, O., et al. 1992. "The excavations in Kebara Cave, Mt Carmel," *Current Anthropology*, 33:497–550.

141 **The oldest artworks in Europe**—See Balter, M. 2006. "Radiocarbon dating's final frontier," *Science*, 313:1560–63; Conard, N. 2003. "Palaeolithic ivory sculptures from southwestern Germany and the origins of figurative art," *Nature*, 426:830–32; Valladas, H., and Clottes, J. 2003. "Style, Chauvet and radiocarbon," *Antiquity*, 77 (295): 142–45; Valladas, H., et al. 2001. "Palaeolithic paintings: Evolution of prehistoric cave art," *Nature*, 413:479; and J.-M. Chauvet, E. B. Deschamps, and E. Hillaire, *Dawn of Art: The Chauvet Cave* (New York: Harry N. Abrams, 1996).

142 **Ochre was mined over 300,000 years ago**—Barham, L. S. 1998. "Possible early pigment use in south-central Africa," *Current Anthropology*, 39:703–10; Barham, L. S., and Smart, P. 1996. "Early date for the Middle Stone Age of central Zambia," *Journal of Human Evolution*, 30:287–90; McBrearty, S. 2000.

"Identifying the Acheulian to Middle Stone Age transition in the Kapthurin Formation, Baringo, Kenya," *Journal of Human Evolution*, 38: A20; Deino, A., and McBrearty, S. 2002. "40Ar/39Ar chronology for the Kapthurin Formation, Baringo, Kenya," *Journal of Human Evolution*, 42:185–210.

142 **ochre also has some medicinal value**—Watts, I. 2002. "Ochre in the Middle Stone Age of southern Africa: Ritualised display or hide preservative?" *South African Archaeological Bulletin*, 57:15–30; see also the summary of ochre uses in Wadley, L. 2001. "What is cultural modernity? A general view and a South African perspective from Rose Cottage Cave," *Cambridge Archaeological Journal*, 11:2 (2001): 201–21, pp. 204–5.

143 **Direct evidence of symbolic ochre use . . . "are not consistent with doodling"**—Henshilwood, C. S., d'Errico, F., and Watts, I. 2009. "Engraved ochres from the Middle Stone Age levels at Blombos Cave, South Africa," *Journal of Human Evolution*, 57:27–47, p. 42.

144 **"The most striking examples"**—Henshilwood, d'Errico, and Watts, "Engraved ochres from the Middle Stone Age levels at Blombos Cave, South Africa," p. 42.

145 **a recent inventory of geometric or nonfigural art**—Von Petzinger, G. 2009. "Making the abstract concrete: The place of geometric signs in French Upper Paleolithic parietal art," *Paleoanthropology Society Meeting Abstracts*, Chicago, March 31 and April 1, 2009, A29; Von Petzinger, G. 2005. "Making the Abstract Concrete: The place of geometric signs in French Upper Paleolithic cave art," MA thesis, University of Victoria, copyright 2009: 1–157.

146 **recent discoveries at Diepkloof Rock Shelter**—Texier, P., et al. 2010. "A Howiesons Poort tradition of engraving ostrich eggshell containers dated to 60,000 years ago at Diepkloof Rock Shelter, South Africa," *Proceedings of the National Academy of Sciences, USA*, 107 (14): 6180–85; Rigaud, J.-P., et al. 2006. "Le mobilier Stillbay et Howiesons Poort de l'abri Diepkloof." (English title "South African Middle Stone Age chronology: New exca-

vations at Diepkloof Rock Shelter: Preliminary results"), *Comptes Rendus Paleovol*, 5:839–49; J. Parkington, et al., "From Tools to Symbols: The Behavioural Context of Intentionally Marked Ostrich Eggshell from Diepkloof, Western Cape," in F. d'Errico and L. Backwell, eds., *From Tools to Symbols: From Early Hominids to Modern Humans* (Johannesburg: Witwatersrand University Press, 2005), pp. 475–91.

147 **objects of personal adornment**—F. d'Errico and M. Vanhaeren, "Earliest Personal Ornaments and Their Significance for the Origin of Language Debate," in R. Botha and C. Knight, eds., *The Cradle of Human Language* (Oxford: Oxford University Press, 2009), pp. 24–60; Bouzouggar, A., 2007. "82,000-year-old shell beads from North Africa and implications for the origins of modern human behavior," *Proceedings of the National Academy of Sciences, USA*, 104 (24): 9964–69; Vanhaeren, M., et al. 2006. "Middle Paleolithic shell beads in Israel and Algeria," *Science*, 312:1785–88; d'Errico, F., et al. 2005. "*Nassarius kraussianus* shell beads from Blombos Cave: Evidence for symbolic behaviour in the Middle Stone Age," *Journal of Human Evolution*, 48: 3–24; Henshilwood, C. S., et al. 2004. "Middle Stone Age shell beads from South Africa," *Science*, 384:404; Ambrose, S. 1998. "Chronology of the Later Stone Age and food production in East Africa," *Journal of Archaeological Science*, 25:377–92; Mehlman, M. J. 1989. "Late Quaternary Archaeological Sequences in Northern Tanzania." PhD dissertation, University of Illinois, Urbana. For a discussion of the significance of personal adornments, see S. Kuhn and M. Stiner, "Body Ornamentation as Information Technology: Towards an Understanding of the Significance of Early Beads," in Mellars, Boyle, Bar-Yosef, and Stringer, eds., *Rethinking the Human Revolution: New Behavioural and Biological Perspectives on the Origin and Dispersal of Modern Humans*, pp. 45–54.

148 **"one of the most fascinating cultural experiments"**—F. d'Errico quoted in Anonymous, 2009. "80,000 year old shells point

to earliest cultural trend." Press release, European Science Foundation, August 27.

149 **If using bone to make tools makes you modern**—d'Errico, F., and Backwell, L. 2009. "Assessing the function of early hominin bone tools," *Journal of Archaeological Science*, 36:1764–73; Backwell, L., and d'Errico, F. 2001. "Evidence of termite foraging by Swartkrans early hominids," *Proceedings of the National Academy of Sciences, USA*, 98 (4): 1358–63; C. K. Brain and P. Shipman, "Bone Tools from Swartkrans," in C. K. Brain, ed., *Swartkrans: A Cave's Chronicle of Early Man* (Pretoria: Transvaal Museum Monograph No. 8), pp. 195–215; P. Shipman, "Altered Bones from Olduvai Gorge, Tanzania: Techniques, Problems, and Implications of Their Recognition," in R. Bonnichsen and M. Sorg, eds., *Bone Modifications* (Orono, ME: Center for the Study of Early Man, 1989), pp. 317–34.

150 **Hunting large and dangerous animals**—Dominguez-Rodrigo, et al. 2005. "Cutmarked bones from Pliocene archaeological sites at Gona, Afar, Ethiopia: Implications for the function of the world's oldest stone tools," *Journal of Human Evolution*, 48 (2): 109–21; Dominguez-Rodrigo, M. 2001. "Hunting and scavenging in early hominids: The state of the debate," *Journal of World Prehistory*, 16:1–56; Dominguez-Rodrigo, M., and Pickering, T. R. 2003. "Early hominid hunting and scavenging: A zooarchaeological review," *Evolutionary Anthropology*, 12:275–82.

151 **exploiting a wider range of animal and plant resources**—See Braun, D. R., et al. 2010. "Early hominin diet included diverse terrestrial and aquatic animals 1.95 Ma in East Turkana, Kenya," *Proceedings of the National Academy of Sciences, USA*, 107:10,002–7; Joordens, J. C. A., et al. 2009. "Relevance of aquatic environments for hominins: A case study from Trinil (Java, Indonesia)," *Journal of Human Evolution*, 57:656–71; Fernandez-Jalvo, Y., Andrews, P., and Denys C. 1999. "Cut marks on small mammals at Olduvai Gorge Bed I," *Journal of Human Evolution*, 36:587–89; Stiner, M. C., et al. 1999. "Paleolithic population growth pulses evidenced

by small animal exploitation," *Science*, 283:190–94; Stiner, M. C., Munro, N. D., and Surovell, T. A. 2000. "The tortoise and the hare: Small-game use, the broad spectrum revolution, and paleolithic demography," *Current Anthropology*, 41:39–74.

151 **Controlling fire is another modern behavior**—Goren-Inbar, N., et al. 2004. "Evidence of hominin control of fire at Gesher Benot Ya'aqov, Israel," *Science*, 304:725–27.

152 **McBrearty and Brooks suggest**—McBrearty, S., and Brooks, A. 2000. "The revolution that wasn't: A new interpretation of the origin of modern human behavior," *Journal of Human Evolution*, 39:453–563, p. 531.

153 **"As I see it, these are the challenges"**—S. McBrearty, "Down with the Revolution," in Mellars, Boyle, Bar-Yosef, and Stringer, eds., *Rethinking the Human Revolution: New Behavioral and Biological Perspectives on the Origins and Dispersal of Modern Humans*, p. 139.

154 **the apparent convergence of behaviors**—Mellars, P. 2006. "A new radiocarbon revolution and the dispersal of modern humans in Eurasia," *Nature*, 439:931–35.

Chapter 10

164 **The overwhelming majority of prehistoric art**—M. Conkey, personal communication to the author. See also, e.g., A. Leroi-Gourhan, *Treasures of Prehistoric Art* (New York: Harry Abrams, 1965); P. G. Bahn, and J. Vertut, *Journey Through the Ice Age* (Berkeley: University of California Press, 1997); R. White, *Prehistoric Art: The Symbolic Journey of Humankind* (New York: Harry Abrams, 2003), p. 6; Bahn, *Cave Art: A Guide to the Decorated Ice Age Caves of Europe* (London: Frances Lincoln, 2007); J. Clottes, *Cave Art* (London: Phaidon Press, 2010); and D. R. Guthrie, *The Nature of Paleolithic Art* (Chicago: University of Chicago Press, 2006).

165 **a single instance of a stone tablet from Spain**—Utrilla, P., et

al. 2009. "A palaeolithic map from 13,660 calBP: Engraved stone blocks from the Late Magdalenian in Abauntz Cave (Navarra, Spain)," *Journal of Human Evolution*, 57:99–111.

166 **a 58,000-year-old factory for processing ochre**—Wadley, L. 2010. "Cemented ash as a receptacle or work surface for ochre powder production at Sibudu, South Africa, 58,000 years ago," *Journal of Archaeological Science*, 37:2397–2406.

168 **the acoustic resonance of painted caves**—Reznikoff, I., and Dauvois, 1988. "La dimension sonore des grottes ornées," *Bulletin de la Société Prehistorique Française*, 85:238–46; I. Reznikoff, "Prehistoric Paintings, Sound and Rocks," in E. Hickmann, A. Kilmer, and R. Eichmann, eds., *Studien zur Musikarchäologie*, Vol. III [Orient-Archäologie], pp. 39–56. Papers from the 2nd Symposium of the International Study Group on Music Archaeology, Monastery Michaelstein, Rahden, Germany, September 2000.

169 **"because of the resonance"**—Reznikoff, I., 2004–5. "On primitive elements of musical meaning," *Journal of Music and Meaning*, 3, sect. 2, 1, accessed June 19, 2010, at www.musicandmeaning.net/issues/showArticle.php?artID=3.2.

169 **studied acoustic resonance at Font-de-Gaume**—Waller, S. J. 1993. "Sound and rock art," *Nature*, 363:501.

169 **flutes dated to more than 35,000 years old**—S. Münzel, F. Seeberger, and W. Hein, "The Geissenkloesterle [sic] Flute—Discovery, Experiments, Reconstruction," in Hickman, Kilmer, and Eichmann, eds., *Studien zur in Musikachäologie*, Vol. III [Orient-Archäologie] pp. 107–18; Conard, N. J., Malina, M., and Münzel, S. C. 2009. "New flutes document the earliest musical tradition in southwestern Germany," *Nature*, 460:737–40.

Chapter 11

174 **"What enables human beings to create and use"**—Tomasello, M. 1999. "The human adaptation for culture." *Annual Review of Anthropology*, 28:509–29, pp. 511–12.

176 **The bonobo language experiment**—Information here is taken from Savage-Rumbaugh and Lewin, *Kanzi: The Ape at the Brink of the Human Mind*; Savage-Rumbaugh, E. S., Rumbaugh, D. M. and McDonald, K. 1985. "Language learning in two species of apes," *Neuroscience and Biobehavioral Reviews*, 9:653–65; and Savage-Rumbaugh, E. S., et al. 1986. "Spontaneous symbol acquisition and communicative use by pygmy chimpanzees (*Pan paniscus*)," *Journal of Experimental Psychology: General*, 115:211–35.

177 **The Clever Hans phenomenon**—O. Pfungst, trans. C. L. Rahn, *Clever Hans (The Horse of Mr. von Osten): A Contribution to Experimental Animal and Human Psychology* (1907; New York: Henry Holt, 1911).

181 **in Bonobos comprehension precedes production**—D. Rumbaugh and S. Savage-Rumbaugh, "Language and Animal Competencies," in N. J. Smelser, and P. B. Bates, eds., *International Encyclopedia of the Social and Behavioral Sciences* (London: Pergamon Press, 2001), p. 8281.

181 **Kanzi's performance in language production**—W. Fields, personal communication to author; P. Raffaele, "Speaking Bonobo," *Smithsonian Magazine* (November 2006).

181 **toddlers have a vocabulary**—Fenson, L., et al. 1994. "Variability in early communicative development," *Monographs of the Society for Research in Child Development*, serial No. 242: 59 (5).

181 **the average high school graduate has a vocabulary**—Hauser, M. D., Chomsky, N., and Fitch, W. T. 2002. "The faculty of language: What is it, who has it, and how did it evolve?" *Science*, 298:1567–79.

183 **Vicki Hearne was an extraordinary animal trainer**—See esp. V. Hearne, *Adam's Task: Calling Animals by Name* (New York: Alfred A. Knopf, 1986), and V. Hearne, *Animal Happiness* (New York: Perennial Books, 1995).

183 **Chimpanzees are rarely observed**—King, B. J. 1991. "Social information transfer in monkeys, apes, and hominids," *American Journal of Physical Anthropology*, 34 (513): 97–115.

184 **"the chimpanzee mothers' attitude"**—Inoue-Nakamura, N., and Matsuzawa, T. 1997. "Development of stone tool use by wild chimpanzees *(Pan troglodytes),"* *Journal of Comparative Psychology,* III (2): 159–73, p. 172.

184 **claims bonobos teach each other**—S. Savage-Rumbaugh, personal communication to the author.

184 **Barbara J. King . . . has analyzed ape communication**—B. J. King. *The Dynamic Dance: Normal Communication in African Great Apes* (Cambridge, MA: Harvard University Press, 2004).

184 **And Jill Pruetz**—J. Pruetz, personal communication to the author.

187 **"I stood still, my whole attention fixed"**—H. Keller, *The Story of My Life: The Restored Classic, Complete and Unabridged* (1904; New York: W. W. Norton, 2003), p. 15.

Chapter 12

191 **V. Gordon Childe proposed in 1936**—See V. G. Childe, *Man Makes Himself* (1936; Oxford: Oxford University Press, 1962).

194 **there is a paradox involved in domesticating plants**—Webster, D. "Backward bottlenecks: Ancient teosinte/maize selection," *Current Anthropology* (in press).

199 **Domestication happened not once in one place**—See Zeder, M. A., et al., eds., *Documenting Domestication: New Genetic and Archeological Paradigms* (Berkeley: University of California Press, 2006); Diamond, J., 2002. "Evolution, consequences and future of plant of animal domestication," *Nature,* 418:700–707; and Balter, M. 2007. "Seeking agriculture's ancient roots." *Science,* 316:1830–35.

200 **"The first revolution"**—V. Childe, *Man Makes Himself,* p. 59.

201 **domesticating silver foxes**—Trut, L. 1999. "Early canid domestication: The farm-fox experiment,"*American Scientist,* 87 (2): 160–70; Lindberg, J., et al. 2005. "Selection for tameness has changed brain gene expression in silver foxes," *Current Biol-*

ogy, 15:R915–R916; Kukekova, A. V., et al. 2008. "Measurement of segregating behaviors in experimental silver fox pedigrees," *Current Biology*, 38 (2): 185–94.

202 **Rye is currently the oldest known**—Hillman, G. C., Legge, A. J., and Rowle-Conwy, P. A., 1997. "On the charred seeds from Epipalaeolithic Abu Hureyra: Food or fuel?" *Current Anthropology*, 38 (4): 648–59.

203 **"The transition from hunters and gatherers"**—B. D. Smith, quoted in Pringle, H., 1998. "The Slow Birth of Agriculture," *Science*, 282:1446.

Chapter 13

206 **when you compare the economics of dealing with a wild wolf**—Data on wolves come from Dewey, T., and Smith, J., 2002. "*Canis lupus*" (online), Animal Diversity Web, accessed June 21, 2010, at animaldiversity.ummz.umich.edu/site/accounts/information/Canis_lupus.html. Data on goats from Carew, B. A. R., et al. "FAO Report on African Sheep and Goats: The Potential of Browse Plants in the Nutrition of Small Ruminants in the Humid Forest and Derived Savanna Zones of Nigeria," in H. N. Le Houérou, ed., *Browse in Africa: The Current State of Knowledge*. Papers presented at the International Symposium on Browse in Africa, Addis Ababa, April 8–12, 1980, and other submissions (Addis Ababa, Ethiopia: International Livestock Center for Africa).

209 **"You develop a village"**—R. Coppinger, quoted in *The Animal Attraction*, Program 1, "The Pact for Survival: How Wolves Became Dogs," Australian Broadcasting Company, accessed June 19, 2010, at abc.net.au/animals/program1.transcript1.htm. See also R. and L. Coppinger, *Dogs: A Startling New Understanding of Canine Origin, Behavior and Evolution* (New York: Scribner's, 2001).

211 **"migratory wolves"**—Cook, S. J., Norris, D. R., and Theberge, J. B. 1999. "Spatial dynamics of a migratory wolf population in

winter, south-central Ontario (1990–1995)," *Canadian Journal of Zoology*, 77: 1740–50.

211 **a remarkable study**—See Germonpré, et al. 2009. "Fossil dogs and wolves from Palaeolithic sites in Belgium, the Ukraine and Russia: Osteometry, ancient DNA and stable isotopes," *Journal of Archaeological Science*, 36:473–90.

216 **"I was not so surprised . . . But I *was* surprised."**—M. Germonpré, personal communication to the author.

219 **There are other early dog fossils**—Napierala, H., and Herpmann, H.-P. 2010. "A 'new' paleolithic dog from Central Europe," *International Journal of Osteoarchaeology*, n.a. doi. 10.1022/or. 1182; Sablin, M. V., and Khlopachev, G. A. 2002. "The earliest Ice Age dogs: Evidence from Eliseevichi I," *Current Anthropology*, 43 (5): 795–809; Nobis, G. 1979. "Der latest Haushund lebte vor 14,000 Jahren," *Umschau*, 19:610; and Davis, S. J. M., and Valla, F. R. 1978. "Evidence for domestication of the dog 12,000 years ago in the Natufian of Israel," *Nature*, 276:608–10. See also the summary in J. Clutton-Brock, *A Natural History of the Domesticated Mammals* (Cambridge: University Press, 1999), esp. pp. 49–61.

219 **a genomewide examination**—vonHoldt, B. M., et al. 2010. "Genome-wide SNP and haplotype analyses reveal a rich history underlying dog domestication," *Nature*, 464:899–903.

Chapter 14

223 **walking larders**—This term is borrowed from the title of an excellent book edited by J. Clutton-Brock, *The Walking Larder: Patterns of Domestication, Pastoralism, and Predation* (London: Unwin Hyman, 1989).

226 **"a process of increasing mutual dependence"**—Zeder, M. A., et al. 2006. "Documenting domestication: The intersection of genetics and archaeology," *Trends in Genetics*, 22 (3): 139–55, p. 1.

226 **Jean-Denis Vigne . . . and his colleagues**—J. D. Vigne, J. Peters, and D. Helmer, "New archaeological approaches to

trace the first steps of animal domestication: General presentation, reflections and proposals," in J. D. Vigne, J. Peters, and D. Helmer, eds., *First Steps of Animal Domestication: New Archaeozoological Approaches* (Oxford: Oxbow Books, 2005), pp. 1–18.

227 **"Probably for the same reason"**—P. Bahn, personal communication to the author.

227 **"The scarcity of artistic depictions . . . What if dogs"**—A. Pike-Tay, personal communication to the author.

228 **Between 323,070 and 504,600 deer were harvested**— Deer numbers from Pennsylvania Game Commission Web site, accessed June 19, 2010, at www.portal.state.pa.us/portal/server.pt?open=514&objID=625882&mode=2).

228 **a well-known kill site in Burgundy, France**—Olsen, S., 1989. "Solutré: A theoretical approach to the reconstruction of Upper Paleolithic hunting strategies," *Journal of Human Evolution*, 18:295–327.

231 **early domestication could be detected in**—Hesse, B. 1982. "Slaughter patterns and domestication: the beginnings of pastoralism in western Iran," *Man*, n.s. 17 (3): 403–17; See also B. Hesse, *Evidence for Husbandry from the Early Neolithic Site of Ganj Dareh in Western Iran*, PhD dissertation, Columbia University, 1978; Ann Arbor, MI: University Microfilms.

231 **he and Melinda Zeder published a now-classic paper**— Zeder, M., and Hesse, B. 2000. "The initial domestication of goats (*Capra hircus*) in the Zagros Mountains 10,000 years ago," *Science*, 287:2254–57.

231 **Domesticated wheat . . . and domesticated rye**—Hillman, G., et al., 2001. "New evidence of Late glacial cereal cultivation at Abu Hureyra on the Euphrates," *The Holocene*, 11 (4): 383–93; Balter, M. 2007. "Seeking agriculture's ancient roots," *Science*, 316:1830–35.

232 **"Then I realized"**—R. K. Wayne, personal communication to the author.

233 **Wild boar . . . were apparently transported**—A. Matsui, et

al., "Wild Pig? Or Domesticated Boar? An Archaeological View on the Domestication of *Sus scrofa* in Japan," in Vigne, Peters, and Helmer, eds., *The First Steps of Animal Domestication: New Archaeozoological Approaches*, pp. 148–59.

234 **many domesticated pigs were spread**—Dobney, K., Cucchi, T., and Larson, G. 2008. "The pigs of Island Southeast Asia and the Pacific: New evidence for taxonomic status and human-mediated dispersal," *Asian Perspectives*, 47 (1): 59–74.

234 **Camelids like llama and alpaca seem to get larger**—G. L. Mongoni Gõnalons and H. D. Yacobaccio, "The domestication of South American camelids: A view from the south-central Andes," in M. A. Zeder, et al., eds., *Documenting Domestication: New Genetic and Archaeological Paradigms* (Berkeley: University of California Press, 2006), pp. 228–44.

235 **the appearance of domesticated pigs in China**—Yuan, J., and Flad, R. K. 2002. "Pig domestication in Ancient China," *Antiquity*, 76:724–32.

235 **domestication occurred independently in at least**—Larson, G., et al. 2010. "Patterns of East Asian pig domestication, migration, and turnover revealed by modern and ancient DNA," *Proceedings of the National Academy of Sciences, USA*, 107 (17): 7686–91.

236 **work . . . on the domestication of the horse**—Outram, A. K., et al. 2009. "The earliest horse harnessing and milking," *Science*, 323:1332–35; S. Olsen, "Early horse domestication on the Eurasian Steppes," in Zeder, M. A., et al., eds., *Documenting Domestication: New Genetic and Archaeological Paradigms*, pp. 245–69; S. Olsen, "The importance of thong-smoothers at Botai, Kazakhstan," in A. Choyke and L. Bartosiewicz, eds., *Crafting Bone; Skeletal Technologies Through Time and Space* (Oxford: British Archaeological Reports, 2001); and S. Olsen, "The exploitation of horses at Botai, Kazakhstan," in M. Levine, C. Renfrew, and K. Boyles, eds., *Prehistoric Steppe Adaptation and the Horse* (Cambridge: McDonald Institute Monographs, 2003), pp. 83–104.

237 **"A communal hunting strategy"**—S. Olsen, personal communication to the author.

239 **"I doubt that Botai hunters"**—Ibid.

240 **bit wear could be used as evidence**—Anthony, D. W., and Brown, D., 2000, "Eneolithic horse exploitation in the Eurasian steppes: Diet, ritual and riding," *Antiquity*, 74:75–86; Brown, D., and Anthony, D. 1998. "Bit wear, horseback riding and the Botai Site in Kazakhstan," *Journal of Archaeological Science*, 25 (4): 331–47.

243 **the "smoking gun" . . . "You can't imagine"**—P. Smith, "Archaeologist chases ancient horse history with adventurous spirit," *Pittsburgh Post-Gazette*, March 25, 2009.

244 **"What's really key here"**—A. Outram, quoted in Anonymous, "Humans farmed horses 5,500 years ago," *Cosmos*, March 6, 2009; accessed June 19, 2010, at www.cosmos.com.

Chapter 15

248 **"Food production could not possibly have arisen"**—Diamond, J. 2002. "Evolution, consequences, and future of plant of animal domestication," *Nature*, 418:700–707, p. 700.

249 **"the ubiquitous tendency of all peoples"**—Ibid.

249 **This old idea was first expressed**—Galton, F. S. 1865. "The first steps towards the domestication of animals," *Transactions of the Ethnological Society, London*, n.s. 1:122–38, reprinted in F. S. Galton, *Inquiries into Human Faculty* (London: J. M. Dent, 1907). Compare the discussion with J. Clutton-Brock, *A Natural History of Domesticated Animals* (Cambridge: University Press, 1995), pp. 8–9, and Diamond, J. 2002. "Evolution, consequences, and future of plant of animal domestication," *Nature*, 418:700–707, p. 702.

250 **a "Secondary Products Revolution"**—Sherratt, A. 1983. "The Secondary Products Revolution of animals in the Old World," *World Archaeology*, 15:90–104; A. Sherratt, "Plough and

pastoralism: Aspects of the Secondary Products Revolution," in I. Hodder, G. Isaac, and N. Hammond, eds. *Pattern of the Past* (Cambridge: University Press, 1981).

251 **who has recently revisited**—Greenfield, H. J. 2010. "The Secondary Products Revolution: The past, the present and the future," *World Archaeology*, 42:1, 29–54, p. 3.

251 **"Successful husbandry requires"**—B. Hesse, "Carnivorous pastoralism: Part of the origins of domestication or a Secondary Product Revolution?" in R. W. Jameson, S. Abouyi, and N. A. Mirau, eds., *Culture and Environment: A Fragile Co-Existence, Proceedings of the 24th Annual Conference of the Archaeological Association of Canada*, Calgary, p. 101.

254 **Large, docile animals offer**—G. R. Bentley, "How did prehistoric women bear 'Man the Hunter'? Reconstructing fertility from the archaeological record," in R. P. Wright, ed., *Gender and Archaeology* (Philadelphia: University of Pennsylvania Press, 1996), pp. 23–51.

257 **The intensive observation of and intimacy with**—See, e.g., Bennet, M. M. 2002. "What's in a name? Berserk Male Syndrome versus Novice Handler Syndrome," *The Camelid Quarterly* (September): 1–3.

258 **the continued functioning of lactase**—W. Durham, *Coevolution: Genes, Culture, and Human Diversity* (Stanford, CA: Stanford University Press, 1991); Tishkoff, S. A., et al. 2007. "Convergent adaptation of human lactase persistence in Africa and Europe," *Nature Genetics*, 39:31–40; Burger, J., et al. 2007. "Absence of the lactase-persistence-associated allele in early Neolithic Europeans," *Proceedings of the National Academy of Sciences, USA*, 104 (10): 3736–41.

Chapter 16

260 **"Horses are swift of foot"**—S. Olsen, quoted in W. S. Weed, "First to ride," *Discover* (March 2002).

261 **Genghis Khan was a Mongul**—Information on Genghis Khan is taken from D. Morgan, *The Mongols*, 2nd edn. (Oxford: Oxford University Press, 2007); Hersh, R. 2009. "The Mongols and Plague: Spreading the Black Death," accessed June 20, 2010, at www.early-middleages.suite101.com/article.cfm/the_mongols_and_plague#ixzz0rhqZ4g2L; and The Applied History Research Group, 1998. "The Islamic World to 1600: The Black Death." University of Calgary, accessed June 22, 2010, at www.ucalgary.ca/applied_history/tutor/islam/mongols/blackDeath.html.

262 **Pax Mongolica**—See J. Abu-Lughod, *Before European Hegemony: The world system* A.D. *1250–1350* (New York: Oxford University Press, 1989).

263 **"as though arrows were raining"**—G. de' Mussi, quoted in Wheelis, M. 2002. "Biological warfare at the 1346 siege of Caffa," *Emerging infectious diseases*, September 5: 8, accessed May 5, 2010, at www.cdc.gov/ncidod/EID/vol8no9/01-0536.htm; R. Horrox, ed., *The Black Death* (Manchester: Manchester University Press, 1994), trans. of G. de' Mussi, pp. 14–26; M. Wheelis, "Biological Warfare Before 1914," in E. Geissler, J. Moon, and C. Ev, eds., *Biological and Toxin Weapons: Research, Development and Use from the Middle Ages to 1945* (London: Oxford University Press, 1999), pp. 8–34.

263 **The few who escaped Caffa**—Original account by M. Platiensis, 1357, quoted in J. Nohl, *The Black Death* (London: George Allen & Unwin, 1926), pp. 18–20.

263 **Death tolls were enormous**—See M. W. Dols, *The Black Death in the Middle East* (Princeton: Princeton University Press, 1977), pp. 39, 67; and J. Kelly, *The Great Mortality: An Intimate History of the Black Death, the Most Devastating Plague of All Time* (New York: HarperCollins, 2005), pp. 11–13.

263 **Untreated plague . . . has a mortality rate of 60–100 percent**—World Health Organization, Fact Sheet No267, revised February 2005; S. Lister, "Untreated pneumonic plague has high mortality," *The Times* (London), August 4, 2000.

264 **another zoonosis: the Spanish flu**—Information on the Spanish flu is taken from Patterson, K. D., and Pyle, G. F. 1991. "The geography and mortality of the 1918 influenza pandemic," *Bulletin of the History of Medicine*, 65 (1): 4–21; Taubenberger, J. K., and Morens, D. M. 2006. "1918 influenza: The mother of all pandemics," *Emerging infectious diseases*, January, accessed May 9, 2009, at www.cdc.gov/ ncidod/EID/vol12no01/05-0979.htm; Johnson, N. P. and Mueller, J. 2002. "Updating the accounts: Global mortality of the 1918–1920 'Spanish' influenza pandemic," *Bulletin of the History of Medicine*, 76 (1): 105–15; and S. Knobler, et al., eds., "1: The story of influenza," *The Threat of Pandemic Influenza: Are We Ready? Workshop Summary* (Washington, DC: National Academies Press, 2005).

265 **Over half of the pathogens**—Woolhouse, M. E. J., and Gowtage-Sequeria, S. 2005. "Prevalence of zoonoses as major human diseases," *Emerging Infectious Diseases*, December, 11, accessed June 10, 2010, at www.cdc.gov/ncidod/EID/ vol11no12/05-0997.htm; Taylor, L. H., Latham, S. M., and Woolhouse, M. E. J. 2001. "Risk factors for human disease emergence," *Philosophical Transactions of the Royal Society of London, B: Biological Sciences*, 356:983–89.

265 **A review of human infectious diseases**—Wolfe, N. D., Dunavan, C. P., and Diamond, J. 2007. "Origins of major human infectious diseases," *Nature*, 447:279–83, p. 282.

267 **as Stephen Budiansky argues**—S. Budiansky, *The Covenant of the Wild: Why Animals Chose Domestication* (New York: Random House, 1994), p. 61.

268 **and humans are encroaching upon areas**—Jones, K. E., et al. 2008. "Disease as a hidden cost of development. Global trends in emerging infectious diseases," *Nature*, 451:983–94.

268 **Which zoonoses successfully make the transition**—Lloyd-Smith, J. O., et al. 2009. "Epidemic dynamics at the human-animal interface," *Science*, 326:1362–67.

268 **Although primates make up only about 0.5 percent—**

Wolfe, Dunavan, and Diamond, "Origins of major human infectious diseases," pp. 279–83.

269 **The density and growth of human populations**—Jones, K. E., et al. 2008. "Disease as a hidden cost of development. Global trends in emerging infectious diseases," *Nature*, 451:983–94, p. 991.

Chapter 17

270 **90 percent of the population of the United States**—Prax, V. 2010. "Number of farmers in US; American family farmers feed 155 people each," accessed June 6, 2010, at www.animal-husbandry.suite101.com/article.cfm/american-family-farmers-feeds-155-people-each-2-americans-farm#ixzz0qBuqHgGI.

270 **40.8 percent of the total land acreage in the United States**—USDA Economic Research Services, 2008.

270 **Ninety-eight percent of all U.S. livestock farms**—USDA Economic Research Services, *U.S. Beef and Cattle Industry: Background Statistics and Information*, 2007.

270 **a similar percentage of family-run farms . . . in China**—Z. Hu and D. Zhang, "China's Pasture Resources," in J. Suttie and S. Reynolds, eds., *Transhuman Grazing Systems in Temperate Asia*, FAO Plant Production and Protection, Series 31, Review 1.

271 **Annual expenditures on pets are huge**—Anonymous, 2009. *Pet Industry Statistics and Trends,* American Pet Products Manufacturers Association (APPMA); Australian Companion Animal Council 2006; Pet Food Manufacturers' Association UK, 2008; M. Fackler, "Japan, home of the cute and inbred dog," *New York Times,* December 28, 2006; J. Chaney, "As Chinese wealth rises, pets take a higher place," Reuters, March 17, 2008.

271 **Pets are not only expensive**—Information taken from Australian Companion Animal Council, 2006; Pet Food Manufacturers' Association UK, 2008; Blue Cross Pet Census of 2007, accessed June 6, 2010, www.bluecross.org.uk; Grundy, E., Stuchbury, R.,

and Young, H. 2010. "Households and families: Implications of changing census definitions for analysis of the ONS Longitudinal Study," *Population Trends*, 139 (1): 64–69; Fackler, "Japan, home of the cute and inbred dog" and Chaney, "As Chinese wealth rises, pets take a higher place."

272 **and regard their animals as family members**—Anonymous, 2007. "Pets are 'members of the family' and two-thirds of pet owners buy their pets holiday presents," Survey No. 120, The Harris Poll®, 2007; Pet Census, 2007, www.bluecross.org.uk; Jones, J. M. 2007. "Companionship and love of animals drive pet ownership," accessed on June 15, 2010, at www.gallup.com.

272 **Pets . . . influence and improve human health**—There is an extensive literature on this topic. See, e.g., A. Fine, ed., *Handbook on Animal-Assisted Therapy. 2nd edn. Theoretical Foundations and Guidelines for Practice* (New York: Academic Press, 2006); C. Chandler, *Animal-Assisted Therapy in Counseling* (London: Routledge, 2005); Headey, B., et al. 2002. "Pet ownership is good for your health and saves public expenditure too: Australian and German longitudinal evidence," *Australian Social Monitor*, 5 (4); P. Salotto, *Pet-Assisted Therapy* (Norton, MA: D. J. Publications, 2001); Headey, B., and Krause, P. 1999. "Health benefits and potential budget savings due to pets: Australian and German survey results." *Australian Social Monitor*, 2 (2): 37; Beck, A., and Meyers, N. 1996. "Health enhancement and companion animal ownership," *Annual Review of Public Health*, 17:247; Serpell, J. 1991. "Beneficial effects of pet ownership on some aspects of human health and behaviour," *Journal of the Royal Society of Medicine*, 84:717; J. Serpell, "Pet-Keeping and Animal Domestication: A Reappraisal," in Clutton-Brock, ed. *The Walking Larder: Patterns of Domestication, Pastoralism, and Predation*, pp. 10–21; and O. Cusak, *Pets and the Elderly* (London: Routledge, 1984).

272 **Being with pets also raises oxytocin levels**—M. O. Daley, *Made for Each Other* (Cambridge, MA: Da Capo Press, 2009), p. xv.

273 **Richard Louv has proposed**—R. Louv, *The Last Child in the Woods: Saving Our Children from Nature-Deficit Disorder* (Chapel Hill: Algonquin Books, 2005).

276 **"the most important development"**—Diamond, J. 2002. "Evolution, consequences and future of plant and animal domestication," *Nature*, 418:700–707, p. 700.

276 **Temple Grandin and Catherine Johnson have written**—T. Grandin and C. Johnson, *Animals Make Us Human: Creating the Best Life for Animals* (New York: Houghton Mifflin, 2009).

276 **Barbara J. King has explored**—B. J. King, *Being with Animals: Why We Are Obsessed with the Furry, Scaly, Feathered Creatures Who Populate Our World* (New York: Doubleday Religion, 2010); H. Herzog, *Some We Love, Some We Hate, Some We Eat: Why It's So Hard to Think Straight About Animals* (New York: Harper, 2010).

277 **Richard Louv mourns**—R. Louv, *The Last Child in the Woods*.

278 **points raised so effectively**—J. M. Masson, *The Dog Who Couldn't Stop Loving: How Dogs Captured Our Hearts for Thousands of Years* (New York: Harper, 2010).

278 **A recent study in the United Kingdom**—McNicholas, J. 2010. "Pets and older people in residential care," accessed June 20, 2010, at www.bluecross.org/uk.

279 **The lovely story of Pale Male**—M. Winn, *Red-tails in Love: A Wildlife Drama in New York* (New York: Vintage Books, 1999); F. Lilien, *Pale Male*, Documentary aired on NATURE—WNET, 2004.

Illustration Credits

1. (left) © 2001 Tim White; (above) photo by C. K. Brain.
2. © K. Schick and N. Toth.
3. © Sileshi Semaw.
4. © Pat Shipman.
5. (left) and (below) Reprinted from the *Journal of Archaeological Science* (2006), 33 (4). Travis Rayne Pickering and Charles P. Egeland. "Experimental patterns of hammerstone percussion damage on bones: Implications for inferences of carcass processing by humans," pp. 459–69. With permission from Elsevier.
6. Reprinted from the *Journal of Animal Ecology* (2008), 77:173–183. Norman Owen-Smith and M. G. I. Mills. "Predator-prey size relationships in an African large-mammal food web." With permission from John Wiley & Sons.
7. © Pat Shipman.
8. Reprinted from Henry T. Bunn and Ellen Kroll, 1986. "Systematic butchery by Plio-Pleistocene hominids at Olduvai Gorge, Tanzania," *Current Anthropology*, 27:431–52. With permission from University of Chicago Press.
9. © Pat Shipman.
10. © Blaire Van Valkenburgh.
11. © 2001 Lucinda Backwell and Francesco d'Errico.

12. (top) © 2001 Lucinda Backwell and Francesco d'Errico; (left) Reprinted from the *Journal of Human Evolution* (2006), 50 (2). Jackson Njau and Robert Blumenschine, "A diagnosis of crocodile feeding traces on larger mammal bone, with fossil examples from the Plio-Pleistocene Olduvai Basin, Tanzania," pp. 142–62. With permission from Elsevier.

13. Reprinted from *Proceedings of the National Academy of Sciences* (2001), 98 (4). Lucinda Backwell and Francesco d'Errico. "Evidence of termite foraging by Swartkrans early hominids," pp. 1358–63. © 2001 National Academy of Sciences, USA.

14. Reprinted from *Proceedings of The National Academy of Sciences* (2001), 98 (4). Lucinda Backwell and Francesco d'Errico. "Evidence of termite foraging by Swartkrans early hominids," pp. 1358–63. © 2001 National Academy of Sciences, USA.

15. Reprinted from *Proceedings of the National Academy of Sciences* (2007), 104 (9). J. Mercader, H. Barton, J. Gillepie, J. Harris, S. Kuhn, R. Tyler, and C. Boesch. "4,300-year-old chimpanzee sites and the origins of percussive stone technology," pp. 3043–48. © 2007 National Academy of Sciences, USA.

16. © The Great Ape Trust.

17. N. Alperson-Afil, G. Sharon, M. Kislev, Y. Melamed, I. Zohar, S. Ashkenazi, R. Rabinovich, R. Biton, E. Werker, G. Hartman, C. Feibel, and N. Goren-Inbar, 2010. "Spatial Organization of Hominin Activities at Gesher Benot Ya'aqov, Israel." *Science*, 326:1677–80. Reprinted with permission from the American Association for the Advancement of Science.

18. Reprinted from *Proceedings of the National Academy of Sciences* (2002), 99:2455–60. Naama Goren-Inbar, Gonen Sharon, Yoel Melamed, and Mordechai Kislev. "Nuts, nut cracking, and pitted stones at Gesher Benot Ya'Aqov." © 2002 National Academy of Sciences, USA.

19. © 2009 Tim White.

20. © University of Tübingen.

21. © University of Bergen, Norway.

22. After Genevieve Von Petzinger (2010).

23. By permission of Jean-Pierre Texier.

24. Reprinted from *Proceedings of the National Academy of Sciences* (2009), 106 (38). Francesco d'Errico, Marian Vanhaeren, Nick Barton, Abdeljalil Bouzouggar, Henk Mienis, Daniel Richter, Jean-Jacques Hublin, Shannon P. McPherron, and Pierre Lozouet. "Out of Africa: Modern Human Origins Special Feature: Additional evidence on the use of personal ornaments in the Middle Paleolithic of North Africa," pp. 16051–56. © 2009 National Academy of Sciences, USA.

25. Reprinted from the *Journal of Archaeological Science* (2008), 25. Lucinda Backwell, Francesco d'Errico, and Lyn Wadley. "Middle Stone Age bone tools from the Howiesons Poort layers, Sibidu Cave, South Africa," pp. 1566–80. With permission from Elsevier.

26. © The Bridgeman Art Library.

27. © University of Tübingen.

28. © The Great Ape Trust.

29. Reprinted from the *Journal of Archaeological Science* (2009), 36. M. Germonpré, M. V. Sablin, R. E. Stevens, R. E. Hedges, M. Hofreiter, M. Stiller, and V. R. Despré. "Fossil dogs and wolves from Paleolithic sites in Belgium, the Ukraine and Russia: Osteometry, ancient DNA and stable isotopes," pp. 473–90. With permission from Elsevier.

30. Reprinted from the *Journal of Archaeological Science* (1998), 25 (4). Dorcas Brown and David Anthony. "Bit wear, horseback riding, and the Botai site in Kazakhstan," pp. 331–47. With permission from Elsevier.

31. © Sandra Olsen.

Index

Page numbers in *italics* refer to illustrations.